Electrical Engineering Interviewing for New College Graduates

Joseph Twomey

ISBN: 1721888659
ISBN-13: 978-1721888658

CONTENTS

1 INTRODUCTION

I had some painful, horrible interviews out of college. Interviews where I was probably closer to suffering from a panic attack than getting a job offer. That being said, I learned a great deal from my interview experiences and with practice and study was able to turn things around. After an initial round of remarkably poor interviews, my performances improved to the point that interviewing was no longer something I feared, but something I looked forward to both as a challenge and as an exciting opportunity. Better yet, I started to get job offers!

While in hindsight it is easy to gloss over my initial interview experiences as terrific learning opportunities, many of the lessons learned could have been avoided if I had been better prepared. With fluctuations in the job market and the overall economy, not everyone may be so lucky to have as many real world 'practice' interviews to build up their interviewing skills. This book was written as something that I wish I could give to myself while in college. And now that I'm occasionally on the other side of the interview table, it is something that I wish I could give to new graduates and even some experienced engineers who I see suffering from some of the same mistakes that I made in the past.

What this book is not, is a condensed four year review of everything you should have learned in your studies. If such a thing was possible, it would be hard to justify the time and money required to go through the learning in the first place. So if you somehow managed to get through four years of your education without actually learning anything... this is not going to save you. Assuming you have the B.S'ng skills to do such a thing, however, I'd imagine interviewing should be a piece of cake!

The purpose of this book is to address two major handicaps that many new electrical engineering graduates seem to have during their first job interviews.

1) Emotional Preparation

By their nature, many engineers (even more so for electrical engineers) tend to be introverts. This certainly is not a universal rule, but is common enough to have formed a near universal stereotype. The old joke told in many sales departments, for example:

Q: How can you tell of you are talking to an extroverted engineer?

A: He stares at your shoes while talking to you (as opposed to his own)...

Humans, in general, do not like to be judged, but this is especially true for introverts. Yet this (being judged), is exactly the point of the interview process. The result can be a defensive lock down. By completely shutting down and revealing as little as possible about ourselves, little can be judged! Yet at the same time, by revealing little about ourselves during an interview and displaying all the excitement and enthusiasm of someone going in for a colonoscopy, we leave little chance of getting a job offer.

This is addressed by providing a brief overview of how most electrical engineering interviews are structured (to limit surprises and help you feel more comfortable with a mental image of what to expect) and by listing actionable items that can be used to improve your interview performance.

2) Practical Engineering Technical Questions and Answers

Many new engineering graduates are unprepared for the types of questions typically asked during the technical portion (or portions) of the interview and it is common for industry folk to bash schools for the woeful job they do preparing new engineers to enter the workforce. This may not be a completely fair judgment of the higher education system, but there does seem to be truth to the fact that some graduates struggle to answer even basic technical electrical engineering questions.

Many, if not most university engineering programs consist of what essentially boils down to advanced math and physics classes. These programs tend to be somewhat weak in developing practical engineering skills, and at times seem more geared to preparing a future generation of PhDs who will be working as engineering professors rather than engineers who will be working on real world problems, despite the disproportionate number of students who will take the second path. Unfortunately, the development of pure math ability may take a preference to conceptual understanding. How else can some students who have studied the quantum physics of semiconductor materials struggle to offer a one word description of a transistor or describe how one can be used to turn on an LED?

Much of the advanced math taught in school is rarely, if ever, used by the majority of real world engineers (that is what computers and software are for). The working engineers who develop the technical questions used to evaluate candidates are not likely to ask advanced calculus or quantum physics questions because they never use such skills themselves. They are going to ask about what they know and what they find important from a day to day engineering perspective. So while you may have just spent four plus years toiling through advanced calculus classes, odds are your interviewer has probably all but forgotten such classes as anything other than distant memories (or nightmares).

This is addressed by providing a sample of technical questions that you may expect to see during an entry level electrical engineering interview, some based on questions I have been asked as an interviewee and some that I have used myself as an interviewer. This is by no means an exhaustive encyclopedia of questions. One aspect of electrical engineering that makes it difficult to create such a list is its vast diversity. The technical questions for any given company will be skewed to their specific application of electrical engineering and for the industry they serve. That being said, it is not generally expected that new engineering graduates will have expertise in any one specific field. The general depth and breadth of knowledge from one new graduate to another will be similar. There will be some common ground, therefore, in the types of technical knowledge companies hiring new graduates will be looking for.

The primary goal of the provided questions is to stress the importance of a solid understanding of electrical engineering fundamentals as well as provide some practical, real world engineering problems that commonly show up during entry level engineering interviews. Where appropriate, additional resources are recommended for further study into this type of knowledge, with a focus on application notes and other industry sources that will provide a real world focus to compliment your school text books.

2 TYPES OF INTERVIEWS

Phone Interview

Phone interviews usually involve a short discussion designed to quickly determine if you are a worthwhile candidate to invest more time into with an in-person interview. This is a type of screening interview. The fact that someone calls you to perform a phone interview is a sign that based on your resume and/or recommendations, you look like a potentially promising candidate.

The most challenging aspect of a phone interview can be the surprise element if you are not provided with notice ahead of time (such as by email). One thing you should not do is try to ignore such calls. While you are job hunting, it is not advised to screen unfamiliar numbers (even if that means dealing with the occasional telemarketer). There is no guarantee that a caller will leave a call back number because the caller may have a long list of potential candidates and a limited amount of time. If a phone interviewer is to leave a voice message and callback number for each unavailable candidate in a large pool of candidates, he is guaranteeing himself many interruptions down the road and he may feel satisfied with whatever number of candidates he was able to make immediate contact with as a large enough pool to find the ultimate candidate. Interviewing is a full time job, so always be prepared!

That being said, it is important to put yourself in the right environment and frame of mind when a call comes in. Most interviewers will have no problem waiting on hold for a short period or even accepting a call back a few minutes later to allow you to pull over into a parking lot, find a quiet room to talk, etc. Besides being dangerous, attempting to field a phone interview while driving means the interviewer will not be getting 100% of your attention, and you will not be able to jot down notes. You also risk driving out of service and dropping the call.

Phone interviews are typically performed either by a Human Resources representative or the hiring manager himself, particularly for smaller companies, and they can come at any time. It is not uncommon for a manager to work through a list of potential candidates over the weekend when his schedule is not as burdened by typical weekday work and meetings and he is more likely to reach candidates, who are less likely to be in class.

4

If the call is from a Human Resources representative, it may just be a "does this person exist and have a pulse" type of call, but if it is from the hiring manager or other engineer, do not be surprised to get into some technical conversation, especially about past projects or internships.

While Job Hunting
• Always keep your phone on you and answer it.
• Be awake and sober during typical work hours.
• Always carry a copy of your resume (or have one close by and quickly accessible).
• Always carry a copy of the job description for each position you apply to (or have them close by and quickly accessible).
• Always carry a pen and notebook (or have them close by and quickly accessible).

Never forget that job hunting is a full time job!

Once I am in a quiet room and ready with a pen and notepad, I usually take 10 seconds to get my composure, think positive thoughts, and make a big goofy smile to help put myself in a good mood.

Introductory Interview

Some companies may invite you in for an 'introductory' interview. This will typically be a half day (or less) interview, but may be extended to a full day hiring interview, as described below. The primary purpose of such interviews is typically to allow companies to filter out candidates that they don't think will fit the bill without having to devote the resources of a full day interview. The interview itself may be scheduled in the morning so that it can be extended into a full day hiring interview if everything goes well.

For this reason, you should always be prepared for and schedule yourself in for an entire day of interviewing, even if you are only asked to come in for a morning introductory interview. That being said, not all companies make real time decisions on extending introductory interviews to a full hiring interview. This may be especially true if a large number of candidates are interviewing for a position. So do not immediately count yourself out if your introductory interview is not extended into a hiring interview.

Hiring Interview

This is the real shebang. At this point, you have been identified as a promising candidate and have passed any pre-hiring interview screenings.

Hiring Interview 2+

Particularly for larger companies, there may be additional hiring interviews to allow for a greater number of employees to meet with you. This is especially true for very competitive positions where a selected pool of candidates is gradually reduced in size over two or more days of interviewing.

If you have made it this far, you are doing things right!

3 COMMON SEGMENTS OF THE INTERVIEW

Human Resources

The human resources portion of the interview typically leads off the interview process. A human resources representative may ask you to fill out some paper work, and possibly request contact information for references if they have not already been provided. While job searching, you should have at least a few references lined up ahead of time so that you will look prepared and to make sure any references you do provide are not getting surprise calls.

Human resources representatives are famous for asking incredibly generic interview questions. A Google search for common interview questions will likely give you a good overview of what to expect. There are many web pages, books and even cell phone apps devoted to discussing the best response to these types of questions, and so they will not be addressed here. It is a good idea to spend some time researching and thinking about this portion of the interview, but in general, a human resources representative will never get you a an engineering job (though he may block it). This is a good time to smile and let your enthusiasm for the position shine through.

One important aspect of this portion of the interview will be to learn about the company benefits; health care, retirement savings, etc. In the tension of the moment, these things can seem insignificant, but are very important to consider in your ultimate decision to accept a position, especially in the United States where health care is typically provided by employers.

While salary negotiation is not the subject of this book, the issue may arise during this part of the interview, with the human resources representative asking about your salary requirements or expectations. It is generally advised to defer any such discussion until after the interview, when (as a desired hire!) you will be in a stronger position to negotiate and you will have had some time to digest and think about the various aspects of the job and the company's benefits package.

Giving a number too early, if lower than the company has budgeted for, is sure to get you less than they could be willing to pay. If too high, you may be pricing yourself out of the position before the company has opportunity to see your true value.

Tour

At some point during the interview, it is common for a company representative to give you a tour of the facility. This will typically be another engineer or manager who is familiar with the job function you will be performing.

This is a great opportunity to get a firsthand view of the work environment. Are engineers packed into a noisy cubical area? Are there quit areas? Is the lab messy, etc.? What types of tools and equipment do they have? Depending on the level of secrecy, you may also get to see some active projects and possibly get a better description of the type of work you could be jumping into if hired.

You tour guide will most likely be doing most of the talking throughout the tour, and generally will not be grilling you for information at this time. This is a good opportunity for you, however, to ask questions and learn as much as you can about the work environment which you could end up spending a significant amount of your time in.

Technical

The most dreaded segment of the interview... the technical portion is where your interviewers attempt to find out about your engineering knowledge.

For recent college graduates, the technical questions will generally be geared towards the fundamentals. It is not expected that you'll have the same type of knowledge as an industry veteran, but you should have a solid foundation from which to grow.

In addition to knowledge, the technical portion of the interview may be used to see how you work under pressure. Do you give up easily or are you driven to work through a tough problem? Do you see difficult problems as a threat or a challenge? Are you proactive about asking for additional information where required or do you shut down? Are you are a quick learner who can apply lessons learned on an early tough problem to another one later on?

The questions listed at the end of this book provide examples of what you might see. How the questions are asked can vary from company to company. In previous interviews, I have been left alone in a room with a printed test, asked to answer verbal questions on a white board while standing in front of a group of engineers, and various other situations in between.

Lunch

For an all-day interview, it is common for one or more of the interviewers to take you out for lunch. While certainly less formal than the office portion of the interview, do not forget that this is still part of the interview process.

It is expected that the interviewer will pay for your meal, but that is not an excuse to order the most expensive thing on the menu. You also should not order alcohol, anything that will be difficult to eat or make you look silly eating, that is messy, or that will give you bad breath. Luckily for engineers, most of your interviewers will care less about whether you use the correct fork to eat your salad (it is doubtful that they know in the first place), but you should still practice good table manners.

When nervous, it is common to either lose all appetite or to become a ravenous pig. A good rule of thumb is to order something that has a relatively small portion so that you will not put yourself into a food comma for the remainder of the interview or look wasteful leaving a plate full of food.

It is also expected that there will be a decent amount of conversation throughout the lunch, so a small portion will give you more opportunity to talk with an empty mouth. Lunch conversation is generally more casual, and is likely a good time for both the interviewer to learn about you and you about him and the company. While it may be common to talk briefly about subjects not directly related to the position or company, this is not the time to be telling funny stories about the times you got really drunk in college.

Remember that you both must eat, and that some silence is totally acceptable. The "this silence is weird, I need to say something" thought that passes through our heads is usually shortly followed by a "d'oh!, why did I say that?". At the same time, you shouldn't be a wet noodle who simply sits in silence the entire time. You do not want to come across as the awkward silent guy who may be uncomfortable to work with.

One of the primary goals for the interviewer is to determine if the interviewee is someone he can see himself working with and getting along with (and vice versa), and lunch is a great opportunity to gather a better feel for one another's personalities. Keeping this in mind, remember to be courteous to the wait staff and thankful for the free meal.

Management

The hiring manager, possibly along with some other managers, will meet with the interviewee at some point during the interview. Typically this portion of the interview is less technical, but this is not always the case if the manager happens to be highly technical.

Managers will likely want to become more familiar with your soft skills and communication skills. If you have worked any internship, they may ask for details about the projects you worked on, how you contributed, any problems or mistakes you have made, etc.

Asking about problems and mistakes you have made or how you would have done things differently in previous projects is a favorite interview question. The one wrong answer is to claim you are a perfect human that has never made a mistake or done anything sub-optimally and that if there have been any issues on projects you have been involved in, they were the fault of others. You do not want to come across as an egomaniac who skirts responsibility or sells out his teammates.

The ability to identify and acknowledge mistakes and areas for improvement is a character trait. It is a sign that you will be the type of employee who will grow and improve over time, continually learning and increasing your value to the company. Everyone, especially those just entering the workforce, makes mistakes. Those who can adapt, recover and learn from their mistakes are the types that managers are on the constant lookout for. It also demonstrates that you are the type of person that others will enjoy working with (or at least will not be running from).

Other questions during this part of the interview may relate to your decision to be an engineer or when you knew wanted to be an engineer. The big no-no here is to claim you heard engineers make decent salaries or something that indicates your career choice is purely practical.

Managers are looking for some passion, so do not be afraid to show it. If asked about your past, this is a good opportunity to talk about your legendary Lego projects as a child or how you got in trouble for dismantling a kitchen appliance to see how it worked. Managers like to see that you are an engineer at heart, and not just by education.

Keeping in mind that interviewing is a two way process, the manager portion of the interview is a great opportunity to get a read on the manager or managers you could potentially be working with. Just as management will be trying to determine if you fit the company's engineering culture, you should be trying to determine if this is the type of manager you would like to work with on daily basis.

Co-Workers

Fitting into the company work culture is important, and so it is common for some of your future potential co-workers to meet with you during the interview process. Even if these employees may not be significantly more experienced than you or hold a position title higher than that which you are applying for, their opinion may still hold weight in the final employment decision.

If potential co-workers do not feel like they would be able to work with you for one reason or another, management will be less likely to add you to the team. Some companies even use interview processes that allow for any employee at any level to veto the hiring of a candidate.

It may seem odd that an employee who is potentially at your same level within the company could have such a major influence in the ultimate decision of your hire, but this is a sign of a company that trusts and empowers its employees.

Company Sell

It is common for many companies to include a 'sell' portion of the interview experience, typically towards the end. At this point, you are most likely exhausted from interacting with many new people throughout day, working your way through tough technical questions, and selling yourself.

A final 'sell' portion of the interview is meant to end on a happier note and leave you with a great image of the company. This may be with a manager higher up in the food chain telling you about what a great company they are to work for and why you should want to join their team.

4 ACTIONABLE ITEMS

Smile

This one 'technique' will easily give you a leg up over a large number of your competition. Opening up and letting out a big smile upon meeting interviewers helps to instantly imprint a favorable first impression in their minds. Studies have shown that humans tend to form opinions about others very quickly (within seconds According to Malcolm Gladwell, in *Blink: The Power of Thinking Without Thinking*), and so a properly timed smile can pay dividends.

And as is commonly repeated, a fake smile quickly turns into a real smile. Just as our emotional state may be reflected by our body language, forcing and controlling our body language helps direct our emotional state (see "*Grin and Bear It: The Influence of Manipulated Positive Facial Expression on the Stress Response,*" published in the journal Psychological Science – among other studies). As an interviewee's emotional state plays such a large role in their performance, taking advantage of this effect can provide a huge benefit.

> A great source of information about body language and how it can be used to your advantage is the Ted Talk: *Your body language shapes who you are,* by Amy Cuddy.

Put things in perspective. It is impossible to cram all four (or more) years of your education into your head the night before your interview, or to instantly acquire the missing experience that may have been in the job description by surfing a few web pages, but you surely can put on a smile with little to no preparation (I hope!). If you are not willing to smile at those you are hoping to work with, do you really want this job?

If you had a business degree and were going in for a job interview, a solid first impression would be expected. Take advantage of the fact that as a new electrical engineering graduate, there is at least one simple thing you can do to give yourself an advantage over some of the competition.

> On one interview, I first introduced myself to a Human Resources representative with a firm handshake, large smile and eye contact. It was as if a switch went off and I could almost see her glowing with delight. She went on to tell me that she thought I would be a good fit in the company (after essentially no conversation) while going on to give me tips for the

upcoming interview and complaining that it seemed many of the other candidates who had interviewed for the position were just too nervous.

As previously mentioned, Human Resource representatives will never land you an engineering job, but they can possibly block you from being hired and so it is still important to leave a favorable impression. In this case, nothing less than a big smile helped me get off on a positive note, feeling confident going into the more technical portions of the interview.

Best of all, she didn't feel the need to subject me to any of the generic HR questions.

Smiling also helps convey a sense of confidence, something that seems to be lacking from many first time interviewees. This is understandable. Getting scared, feeling nervous, losing confidence, etc., is part of life and is a natural reaction to a stressful situation. These feelings can, however, hold us back from our dream job.

The confident, enthusiastic, and energetic applicant is always going to have an advantage. Instead of denying emotions exist, we should acknowledge them, learn to overcome the negative ones (which are commonly irrational to begin with), and use positive emotions to our advantage. Think of this as a mental challenge.

As written by John Milton in *Paradise Lost*, "The mind is its own place, and in itself can make a heaven of hell, a hell of heaven." While it is easy to fall into the trap of seeing an interview as hell, it is in your power to turn it into something more enjoyable, and this can begin with a simple smile!

Think Positive

Negative thoughts can be paralyzing. They can act like a virus that takes over the hosting brain, preventing it from more useful processing and draining energy. This is especially true during the technical portion of the interview. If questions start to become difficult or we start to hit a dead end on a problem, it is easy to start falling into a negative thought process...

'Gosh darn it! I don't know what I'm doing. Gosh darn it! I still don't know what I'm doing. Gosh darn it! I should have studied this subject more. Gosh darn it! I'm not going to get this job...

The last line is interview suicide. As my father would always reply to such statements when I was a child, "If that's what you think, you're right!"

Negativity and self-deprecation can also, unfortunately, be used as a bonding tool. Putting ourselves down may make us more approachable and likeable (or so we think). In reality, this is not a good interview tactic. This sounds completely obvious, and it is, but can be surprisingly easy to do and is probably more common than you'd expect. It may be a reflective response to an embarrassing mistake.

In high school, I was someone who had no problem with AP Calculus, but had great difficulty figuring out change when working my part time job as a cashier.

Customer: "Oh wait, I have some change, here is 51 cents."
Me (cashier): "Sorry, I already typed in the payment amount."
Customer's Facial Expression: 'Are you serious?'
Inside My Head: 'Darn it! Is there a calculator around here!?'
(Much of this probably had to do with the tension of the moment rather than my actual math ability).

Likewise, most engineering interview questions I've had to solve by hand that involved a significant amount of math work were ended by the interviewer before I ever had to do the math drudge-work, usually along the lines of:

"It's obvious you know how to solve this, don't worry about the decimal division let's move to the next problem."

After all, real world engineers use computers and calculators, and so a solid understanding of the fundamentals is what most interviewers are really looking for. But in one interview, the line never came (despite my dramatic pause), and I had to do.... DECIMAL DIVISION. I was struggling a bit and taking longer than I felt I should with something I hadn't done by hand since grade school and sheepishly said,

"I struggle with basic math."

DOPE! Who's going to hire the engineer who proclaims he is deficient in "basic math"!?

Optimism is attractive. People are drawn to it. People, on the other hand, generally do not look forward to bringing in another grump to complain about everything under the sun. One thing that seems to commonly hold back engineering candidates from showing their true positive personality is self-consciousness.

Engineers can be a self-conscious lot, and part of this may be due to their quirky nature. But it should be pointed out that quirkiness among engineers is not just accepted, but possibly expected. If you are too normal, people may start to wonder if you were spending your college nights partying and socializing instead of slaving away over a new Linux install on your computer or trying to get some LEDs to blink in a cool pattern.

I'm certainly not suggesting that you should go into an interview with messy hair, your pants on backwards, and acting like a television sitcom version of a mad scientist. You should not, however, be overly concerned about any odd quirks you may have and allow this to prevent your personality from shining through.

Quirks can be good. They are what make us unique, memorable, and possibly even more approachable than Joe Cool, who manages to act in a completely 'normal' manner at all times (how boring). The best companies know that it is good to have a range and diversity of unique personalities. Some of the strangest people I know are engineers, and these may be some of the people who will be interviewing you down the road. Do not think you need to impress your interviewers by displaying how incredibly normal you are. Nobody gets excited about hiring a normal candidate.

If you are excited by some technology you are shown, express that excitement. If you become curious about projects that are mentioned, ask about them. Especially in larger companies with a great number of candidates, interviews can become repetitive for the interviewers. They may just be spewing out the same speech they've given before and counting the minutes until they can get back to working on something more exciting.

Give them something to get excited about; yourself. This is not the time to disappear. It is not the job of the interviewer to bring out the best in you; you must present it to him. Smile, think positive, and let your guard down and your engineering enthusiasm out!

Think Out Loud

The interviewing process is not a time to be 'living in your head.' This is especially true during the technical portion of the interview. You may be on the verge of a correct answer, but your interviewers will have no clue if you are sitting in silence. The silent stare not only denies your interviewers a chance to know what you know and to get insight into your method of thinking, but denies them a chance to help you with a small hint or guidance.

If the silence is stretched for too long, it starts to make things awkward, and you definitely don't want to be remembered as the awkward guy who was making everybody fidgety. Long periods of silence may also make you feel awkward. Such moments can turn into confidence vacuums, and result in a self-destructive pattern of negative thinking. At the very least, thinking out loud keeps you focused. It forces you to weigh your thoughts more diligently and prevents you from collapsing in on yourself.

It is also not uncommon for your interviewers to provide assistance if you are really struggling or heading down the wrong track. The question itself may be designed in a way that you are not expected to be able to answer it. This could be test to see how you face adversity, or if you can learn on the spot.

If you are really stuck on a technical question, thinking out loud helps organize your thoughts and narrow in on the problem while showing your interviewers what you do understand. "I can see this op-amp is used in a negative feedback configuration, meaning the output will move in such a way to attempt to bring the voltage on the negative input terminal equal to that on the positive input terminal." Making knowledge statements such as this can give you an advantage over other candidates who struggled with the question and may have known the same things, but never expressed it.

This certainly is not a license to BS, however, as this tactic almost never works to do anything other than make you look bad. If a question requires knowledge that you know where to find, but not off hand, simply say it. "I don't have this conversion factor memorized, so I would typically refer to my textbook (or Google) in this situation."

Memorization is considered the lowest form of intelligence. Dogs can memorize words and commands, but no one is going to hire them to do design work. Most experienced engineers should appreciate this, especially as modern technology makes such information so easily attainable. Interviewers will typically be much more interested in your thought process.

In any case, do not give up on technical problems unless they are clearly outside your level of knowledge and understanding:
• Keep pushing.
• Break the problem into smaller components and logically work your way through them out loud.

• Make knowledge statements. Even if you do not know the final answer, allow your interviewers to see what you do know (but DON'T BS). This may lead to some guidance and helpful hints, and may help demonstrate that even if you do not know the answer, you are close.

• Acknowledge pieces of information that you know are holding you back. Assuming some obscure fact is required, simply state this is the type of information you would typically refer to your textbooks or an Internet search for.

• Keep pushing.

When you clearly do not know...

Sometimes, you simply will not know the answer, and no amount of concentration and hard work will get you any closer. In these cases, the best answer may simply be to acknowledge that you don't know the answer, but that you will do some more research into it later. The one thing you do not want to say is "can't". There is a difference between not knowing the answer and not being able to find an answer. Engineers encounter problems they are not immediately able to solve all the time. Part of being an engineer is being able to seek out the missing knowledge required to solve a problem, even if that means asking for help.

In cases where an interviewer provides you with significant help on a problem, make sure to pay attention and understand the help you receive. Don't simply nod your head in an effort to 'look smart'. Another question further down the line may require the same knowledge, and this could be a test of you learning ability.

If the interviewer pushes you on to the next question for scheduling (time) reasons before you are able to complete the question and does not provide you with an answer, be sure to either make a mental note of the question or quickly jot it down on your notepad (or at least the part that stumped you).

There are two reasons for doing this.

First, after the interview, you can use other resources (your textbooks, other engineers, the Internet, etc.) to solve the problem, at which point you should send an email to the respective interviewer (since you will have asked for his card), letting him know the problem has been weighing on your thoughts and you found a solution. This shows some gumption on your part.

Second, it is possible that you may encounter this or a similar question again in future interviews within the same industry. Electrical engineering is such a broad discipline that it is almost impossible to practice every type of problem that you could encounter across the entire spectrum of jobs. Assuming there is one particular part of the industry that you are interested in, however, you are likely to start seeing the same sort of technical questions in interviews.

Ask Questions

Most people love talking about themselves, and engineers are not too much different; they love talking about their projects! So don't be afraid to show some curiosity.

A certain character flaw found in some engineers is fear of appearing wrong or unknowledgeable. Engineers work with their brains and commonly pride themselves on their knowledge. Asking a question, therefore, demonstrates a lack of knowledge to at least one other individual and so is avoided altogether. The question is a flag to the world that this particular engineer is lacking in some way. This type of attitude will severely limit your potential as an engineer (and person in general), and likely will fail to trick other engineers anyways.

There are two major reasons (and probably more) why this attitude is bad for engineers.

The first is that by not asking questions, you are not going to learn as much or as fast as someone who is curious and constantly asking questions.

The second is that certain tasks that require learning may take you significantly longer than someone who is not afraid to seek help when they recognize they are in over their head or are starting to notice that too much time is being wasted (a time sink).

A good manager will recognize curiosity and the willingness to ask questions as a sign of confidence. With ever tighter schedules and increasing project complexity, it is difficult to find the time to devote to training new hires. You will not be expected to be as knowledgeable as an industry expert upon your arrival to your new job, but you will be expected to be self-motivated and driven to learn. Companies want employees who take the initiative to get things done, even if that means asking questions.

One of the greatest strengths that engineers at all levels can have is the ability to learn new things and in the process, continually increase their value to the company. This is also a trait that is difficult to evaluate during a short interview process. Demonstrating curiosity and the courage to ask questions, however, is one strong indicator that a candidate has a 'mental growth' mindset.

In addition, asking questions is your most powerful tool in learning about your potential employer. Just as a company may be evaluating you as a potential employee, you should be evaluating the company as a potential place to work. What does a typical engineering work day look like? What type of work environment do they have? What type of culture and engineering philosophy is present? Can you imagine working with the employees you meet?

Engineering seems to become a way of life, and given the large amount of time that you will be devoting to your future job, it is important to learn as much as you can during the interview. Do not forget that interviewing is a two way process.

Stand Up and Write on a Whiteboard

Most engineering interviews will occur in personal engineering offices and/or conference rooms, both of which will usually contain a whiteboard. Stand up and write on a white board without being asked and it will show confidence and initiative (qualities that are very attractive in entry level engineers).

When interviewing for your first engineering job, it is likely you do not have a large portfolio of previous designs. It is common, therefore, to receive questions about your senior project, the personal project you bring in with you on your interview, or projects you may have worked on during an internship.

You should be prepared to explain these projects and your participation in detail. Practice talking about them in front of fellow students, teachers, or anyone else you can get to sit in front of you. Use a whiteboard or chalkboard to practice your board drawing skills as you talk. A little practice here can go a long way. You are almost guaranteed an opportunity to talk about your previous projects so there is really no excuse not to prepare.

One important note regarding the use of someone's whiteboard; while I highly recommend taking the initiative to use a whiteboard, you should always ask permission before actually marking up the board. Typically boards will be erased before bringing in guests for interviews, but if there is anything on the board, never erase it without permission.

If a whiteboard is not available, you can always use some paper from your notebook for the same purpose, though a board is certainly preferable.

> My first interview breakthrough occurred when a group of interviewers were asking me about my senior robotics project. I loved robotics lab, and had spent virtually every free minute I had working on Rocky and Bullwinkle (the names of the two robots my partner and I designed for our senior year robotics competition). So the situation suddenly changed from an intimidating interview to a fun conversation with the opportunity to brag about our bots.
>
> At one point I was having difficulty explaining a certain mechanical feature of Rocky that an interviewer had asked about, and upon seeing the whiteboard, thought the easiest way to answer his question would be with a quick drawing. Once at the board, I realized it was useful for explaining other design elements of the robots as well. Pretty soon I was giving a sales pitch about how awesome our robots were and the interviewers seemed pretty impressed.
>
> Within a couple of days I received a phone call and my first verbal offer. Since this interview, I've always tried to make it a point to use a whiteboard where possible and when it can facilitate a better response to a question or explanation about a project.

Bring in a Project

Listing hobby projects on your resume and bringing one in for an interview is especially important for new graduates because the lack of differentiating material and experience at this stage in your career. Many graduates competing with you for a given position have taken similar classes, received similar grades, worked on similar senior projects, etc.

Personal projects are a way to differentiate yourself and show that you have a real passion for engineering. They also link you with a memorable identifier in the interviewers' memory. Engineers, in general, like talking about projects, so it can be exciting for an interviewer to see what you have been working on, and maybe learn a thing or two from your experience on the project.

If you have multiple projects that may be relevant to a particular position and that you can't decide between, bring a couple of them in, but don't get too carried away. If you have a suitcase of projects the interview process may get derailed and you will not have enough time to get into in-depth conversation about them. Most interviews are scheduled and planned in advanced and usually don't include a significant amount of time for show and tell (the fact that most candidates won't bring anything in is why you want to). Too many projects can dilute the amount of time you can spend talking about any one of them and prevent you from being able to delve into the details of any specific project (which is where real engineering happens). You can always remind your interviewers to check out your website to see the rest of your personal projects and corresponding documentation.

It's a good idea to thoroughly document the project and put the documentation up on a website. This is useful even for experienced engineers because it allows your potential employer to see examples of your work. Some top tier engineering schools have all their students do something like this (create a website for displaying their projects) by default.

To this day, I bring in a personal project to every interview I attend. If you don't have any hobby projects, START ONE!

Practice!

Practice makes perfect. Interviewing is not something most of us do very often, and it involves skills that generally do not fall under the strong suit of engineers. And while many engineers will try to cram and study for the technical portion of the interview (which should be their strong suit), few will practice for the rest of their interview experience. With all the emotion involved in going to your first real interview, why would you plan on just figuring things out on the spot?

Many schools have Career Services Departments that will offer free interviewing practice. This is a great resource to take advantage of and to receive feedback from an experienced individual. Practicing with friends also provides an opportunity to build up your interviewing memory muscles, and allows you to work out solid replies to common interviewing questions.

And since you are an engineer, take advantage of technology and video record your practice sessions. Use the recording to watch your body language and determine where improvements can be made.

The Essentials

Arrive Early

Arrive at least 10 minutes early for your interview. Many interviews follow a strict schedule, and arriving late will give a bad impression to your interviewers before you even begin. It may also cause them to rush your interview to keep on schedule. There may be some paper work and/or security protocols that you need to complete before starting the interview and so you should provision some time for this as well.

If possible, do a dry run to the place of interview the day before the actual interview. Making the trip at the same time you plan to for your interview is ideal as it will give you an indication of the traffic conditions, which can vary significantly throughout the day. Google Maps also offers route time estimates for specific weekday and time of departure combinations.

Dress Well

As the saying goes, dress to impress. These days, the dress code for most engineering departments is informal, but the fact that employees are not wearing suits does not mean you do not need to wear one to the interview. Wearing a suit shows that you are serious about the job opportunity and displays a level of respect towards those that will be interviewing you. While many place little importance on your dress habits, there are some who may be negatively influenced by your choice to dress down.

This should be another no-brainer, unless you are explicitly instructed not to, wear a suit!

Research the Company

Before going into an interview, you should carefully study the company's website, perform a Google News search about the company, talk to any employees you know who already work there, etc. Having knowledge about the company before the interview demonstrates that you have a real interest in working there. It also allows you to better prepare for the technical portion of the interview. Does this company design low power portable devices, high reliability and safety critical products, or specialize in some other specific area that you can focus your pre-interview study on?

Arrive Prepared

Make sure you have multiple copies of your resume. Ideally, everyone you meet with during the interview process will have already had time to sit and look over your resume before the interview begins. But it is not uncommon for some interviewers to have never seen your resume, or if they have, to have forgotten to print it out or bring it with them to a conference room.

- Have the name and office phone number of your primary interview or Human Resources contact (don't assume the receptionist is expecting you and knows who you will be meeting with).
- Have a folder with a fresh notepad and pen or pencil to take notes, as well as a familiar calculator.
- Have reference information for at least two references.
- Turn your phone off for the interview.

Pump Yourself Up

In the sports world, it is common to perform some pump-up activities before a game, whether by blasting music, giving or listening to some motivational speeches, banging heads together, etc. So why is it so different in the professional world? Given your early arrival, you should have plenty of time to blast your favorite pump up song in the car, read over a list of your greatest assets to remind yourself of what a great engineer you are, do some fist pumps, or whatever it takes to put yourself in a positive mindset.

Collect Business Cards

Throughout the interview process, you should ask for business cards from those you meet with if they do not voluntarily provide them to you. This gives you contact information so that you can write follow up emails after the interview thanking your interviewers for their time, reconfirming your interest in the position, and following up on any unanswered questions that came up during the interview. Keeping business cards out on the table during your interview also helps remind you of the names of those you are speaking with in case you draw a blank.

Collect Interview Questions

As the saying goes, "what doesn't kill you makes you stronger". If nothing else, every interview is an opportunity to hone your interview skills. Whenever an interesting or particularly challenging question comes up, make a mental note and write it down as soon as you get back to your car after the interview. It is a good idea to build yourself an interview study packet, which you can grow over time with new interview questions and well thought out answers.

Keep Things in Perspective

What's the worst that can happen?

You may not get a job for reasons beyond your control. You may have a great interview and the hiring manager may want to hire you, but a sudden budget cut may eliminate funds for the position.

You may have a great interview and impress an insecure manager or other interviewers so much that they feel threatened about the prospect of hiring you, as they see you as a competitive threat!

You may have been the second choice out of dozens of candidates, with the first being some engineering god who has nothing better to do in life but eat, drink and breath Maxwell's Equations (or who happens to be the owner's nephew).

At the very least, the interview will likely be a great learning experience that will help improve your future interview performances.

Judge Your Judges

One important thing to remember throughout your interview process is that the judging goes both ways. You are not just vying for a job, you are evaluating if this is a place that you would like to spend a significant amount of your time and if these are people you can see yourself working with.

It is easy to get caught up in thinking of interviewing as a one way process. But remembering that you are also an interview 'judge' has multiple benefits.

First, it promotes a more confident mindset by mentally placing yourself on a similar level as your interviewers.

Second, it can help you avoid ending up in a job you ultimately are not happy with. Work consumes a significant portion of our lives, and can have a huge impact on our overall happiness. Unless you are in desperate need for employment and plan on using a first job as a stepping stone to something better, you should be reasonably selective about your employment. You don't want to end up in a miserable situation only to look back and realize that you could have and should have seen the warning signs during the interview.

I once had an interviewer at a small company spend the majority of our time together telling me how awesome he was. He was the type of guy who, when in school, would stay late after class and continue to pick his professor's brain (or so I was told). He eventually asked me to solve the transfer function for a very complex analog circuit that he drew from memory. At the time, I doubt I could have solved it in less than a day with a computer and all my books. I couldn't help but get the impression that the point of the whole exercise was to leave me in awe at his brilliance. Needless to say, by the end of the interview I did not come away thinking this was someone I wanted to work with.

5 SAMPLE TECHNICAL QUESTIONS

Hardware Questions

HW_Q1:
What is the voltage at V? in the below circuit?

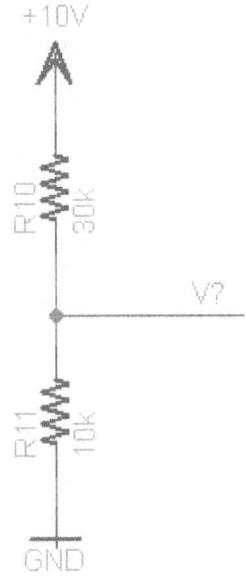

HW_A1:
2.5V

The voltage divider rule is one equation that you (as an electrical engineer) should have memorized for life. It is among the few that you will use time and time again.

Vdivider = Vin (Rlow / (Rlow + Rhigh)) = 10V (10kOhms / (10kOhms + 30kOhms))

HW_Q2:

Assuming 0.1Amps will continuously pass through a 220Ohm resistor for a given application, what is the minimum power rating the resistor must have?

HW_A2:
2.2W (minimum, with further explanation...)

Two equations that should be more familiar to you than your middle name are (for DC circuits):

$V = IR$ (Ohm's law: Voltage = Current x Resistance)

$P = IV$ (Power = Current x Voltage)

Using Ohm's law, we can calculate the voltage across the resistor; 0.1A x 220Ohm = 22V. Knowing both the voltage across the resistor and the current passing through the resistor, we can calculate the dissipated power; 22V x 0.1A = 2.2W. If you have one of the alternate power equations memorized ($P = I^2R$), you can find this answer in one step, but why memorize more equations than necessary?

Hopefully, the numerical answer to this question is a piece of cake. Answering correctly will prevent the interviewer from suddenly remembering they have another meeting to attend to and cut the interview short... But instead of just giving a correct numerical answer, provide some additional insight to demonstrate that you have some real world understanding of the problem, and not just the ability to put numbers into memorized equations.

You may mention that for a real world circuit, continuously dissipating power in a resistor at its maximum power handling capability does not make for a very robust design. You are not likely to find a large selection of 2.2W resistors on the market in any case. 5W resistors are more commonly available and will give the design some margin. It should be mentioned, however, that a more thorough understanding of the application is required to make a final decision, such as at what temperature range the circuit will be exposed to, if it will be used in an enclosed or insulated environment that may limit convection cooling, if it will be placed near other components which may experience their own heating, etc.

Additional comments along these lines will demonstrate that you are already thinking like an engineer, as opposed to a book smart college graduate who is yet able to see past the numbers.

HW_Q3:

What is the voltage at V? in the below circuit?

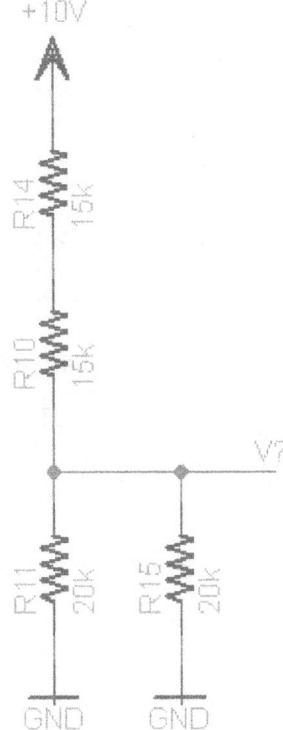

HW_A3:
2.5V

It should be quickly apparent that this circuit can be reduced to an equivalent 2 resistor voltage divider circuit such as used in HW_Q1. The two high side resistors are in series and therefore add together to 30k. The two lower resistors are in parallel and the equivalent resistance, therefore, is the product over sum for those resistors (20k x 20k) / (20k + 20k) = 10k (remember that the product over sum calculation only works when evaluating two parallel resistors!).

Series and Parallel Components

You should know that resistors and inductors sum in series and capacitors sum in parallel. It is not uncommon to see similar questions with more complicated component configurations testing to see your knowledge of how configurations of common passive components can be reduced to a smaller number of components.

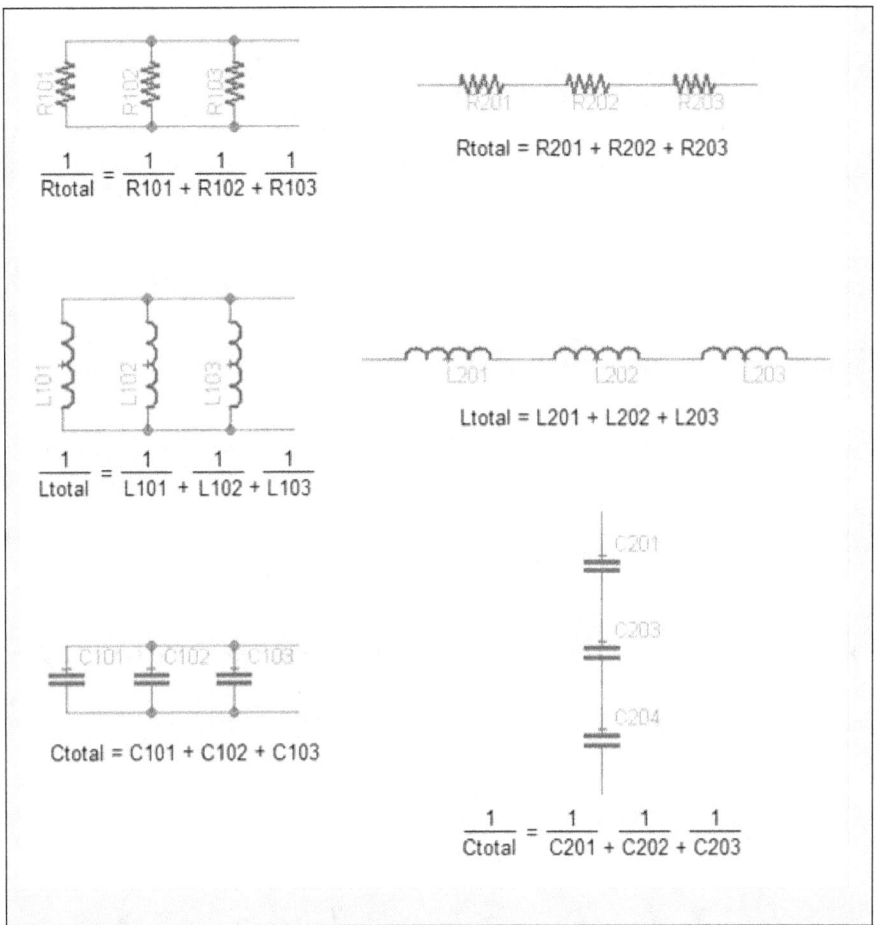

$$\frac{1}{Rtotal} = \frac{1}{R101} + \frac{1}{R102} + \frac{1}{R103}$$

$$Rtotal = R201 + R202 + R203$$

$$\frac{1}{Ltotal} = \frac{1}{L101} + \frac{1}{L102} + \frac{1}{L103}$$

$$Ltotal = L201 + L202 + L203$$

$$Ctotal = C101 + C102 + C103$$

$$\frac{1}{Ctotal} = \frac{1}{C201} + \frac{1}{C202} + \frac{1}{C203}$$

Advanced Network Analysis

While advanced network analysis questions may pop up on the occasional interview, they are not very likely. As an engineer, you will not spend much time examining overly complex circuit networks, trying to figure out the voltage at some particular node or the current through some particular component (and if you need to, you will have a computer). Many such problems that we had to work through in school seemed to have been contrived for the sake of analysis and math work.

Two names you should be familiar with related to circuit analysis, however, are Kirchoff and Thévenin.

Kirchhoff's voltage law (KVL) and Kirchhoff's current law (KCL) are fairly intuitive and well understood by the majority of us familiar with basic electronics.

KVL: The voltages around a closed path in a circuit must sum to zero.

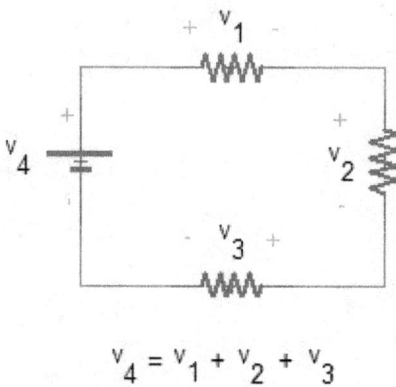

$$v_4 = v_1 + v_2 + v_3$$

KCL: The sum of the currents entering a node must equal the sum of the currents exiting a node.

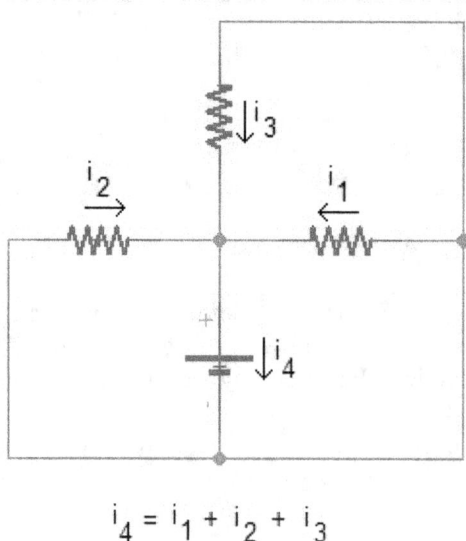

$$i_4 = i_1 + i_2 + i_3$$

Thévenin's theorem states that it is possible to simplify a linear circuit to a single voltage source and resistance connected to a load, to allow analysis of the effect of different loads.

The Thévenin voltage is found by removing any load resistance from the load connection points and calculating the resultant voltage at those points.

The Thévenin resistance is found by shorting any and all voltage sources and opening any and all current sources and calculating the total resistance between the load (open) connection points.

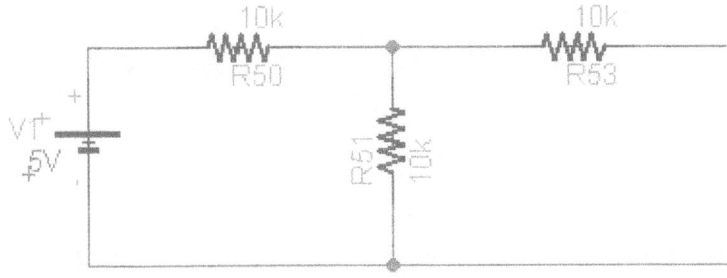

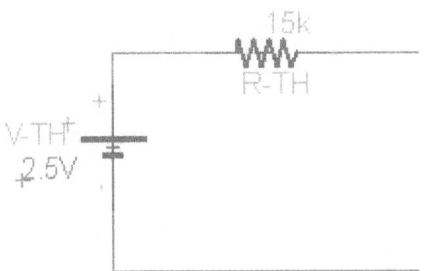

In the above example, the Thévenin voltage is found with the voltage divider equation on R50 and R51. Assuming no load, no current flows through R53 (and therefore no voltage drop).

The Thévenin resistance is found by the parallel combination of R50 and R51 (after shorting the V1 voltage supply) in series combination with R53.

Reducing a circuit to its Thévenin voltage and resistance is sometimes referred to as Thévenizing.

HW_Q4:

What does this schematic symbol represent?

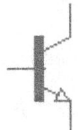

HW_A4:

A transistor; specifically, an NPN BJT (Bipolar Junction Transistor).

Piece Parts

While every new graduate should be intimately familiar with the transistor, there are some components that may come up during an interview that you may not have run across on a regular basis at school.

One common interview strategy is to show a candidate one of the company's actual schematics, and ask him to describe everything he can about it. This allows the interviewer to get an understanding of what type of circuits the candidate is familiar with, as well as to ensure he is familiar with common component symbols.

Listed below are some common components you should be familiar with that may not have been discussed much in school.

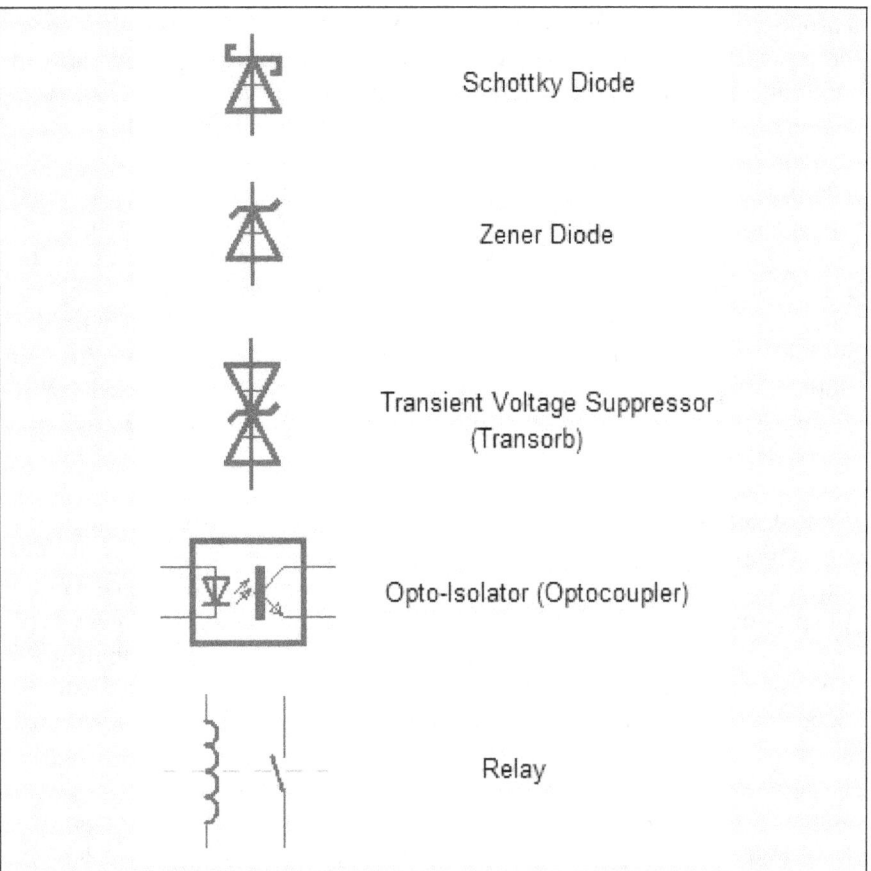

Schottky Diode

Zener Diode

Transient Voltage Suppressor
(Transorb)

Opto-Isolator (Optocoupler)

Relay

Schottky Diode - The Schottky diode is essentially a very good diode. It has a lower voltage drop than a traditional diode (~0.3V instead of ~0.65V) and has a faster reverse recovery time. This makes Schottky diodes ideal for use in switching power supplies or for rectifying large currents.

Zener Diode - The Zener diode is designed with a precise "breakdown" or "avalanche" voltage which can be used to generate a voltage reference, low current voltage source, or to protect an IC pin from seeing too high a voltage.

Transient Voltage Suppressor - The Transient Voltage Suppressor (TVS), or Transorb, is a protection device that is used to clamp voltage over a specified value during Electrostatic Discharge (ESD) events. These devices can be either bidirectional or unidirectional (bidirectional symbol pictured above) and can be thought of as special purpose Zener diodes that are designed to respond quickly to voltage spikes and handle very large peak current surges, dissipating hundreds of Watts for a brief moment. These devices are important for real world products which, when plugged into an AC outlet, may be exposed to voltage spikes from other noisy electronics on the same line or from other electrical disturbances (such as from lighting). Even battery powered electronics handled by humans will commonly use TVS protection because humans have a tendency to build up potentially damaging levels of static charge. It is not uncommon to build up thousands of volts of potential, and while we are all aware of the potentially painful shocks associated with large static discharges, we may discharge up to a few thousand volts without even noticing. For this reason, it is common for electronics manufacturing companies to implement policies to reduce the possibility of ESD damage, such as by requiring technicians and engineers handling unprotected PCBs to wear grounded wrist straps.

Opto-Isolator - Opto-Isolators, or optocouplers, are used to transmit a signal between two electrically isolated systems (e.g there is no electric current flow between the two systems). The transmitted signal is converted to an optical signal (with an LED) on the transmitting side and converted back to an electrical signal on the receiving side (with a phototransistor). These are one way devices, meaning at least two must be used for bi-directional communication (although the two may be in the same package).

Isolated communication is commonly used to transfer information between systems with different or potentially different voltage rails. This is common in industrial environments, where different ground potentials can lead to large and potentially destructive current flows between systems. Isolation is also common in safety critical applications, either to protect users from high voltages or as part of a redundant system that uses multiple isolated power supplies. Such redundant power supplies are kept isolated from each other so that a single short circuit fault cannot bring them all down together.

Recently, other signal isolation technology has begun to become popular as well, and small isolation chips are now available that use inductive and capacitive means of communication.

You will commonly hear electrical isolation referred to as "galvanic isolation."

Relays - Relays are electromechanical* switches that most engineering graduates are familiar with, although there is occasionally confusion about the schematic symbols and naming convention of common relays, which are discussed below.

*solid state relays are the exception as they do not have any moving parts.

Throwing Poles

The terminology used for relays and switches can be a bit confusing to those not familiar with it. Relays are typically described in terms of poles and throws. The pole can be thought of as the actual switching mechanism, so a double pole device has two separate switches. The throw indicates whether the pole simply opens and closes a single line, or if it switches a common line between two or more lines.

Shown below are common relay types.

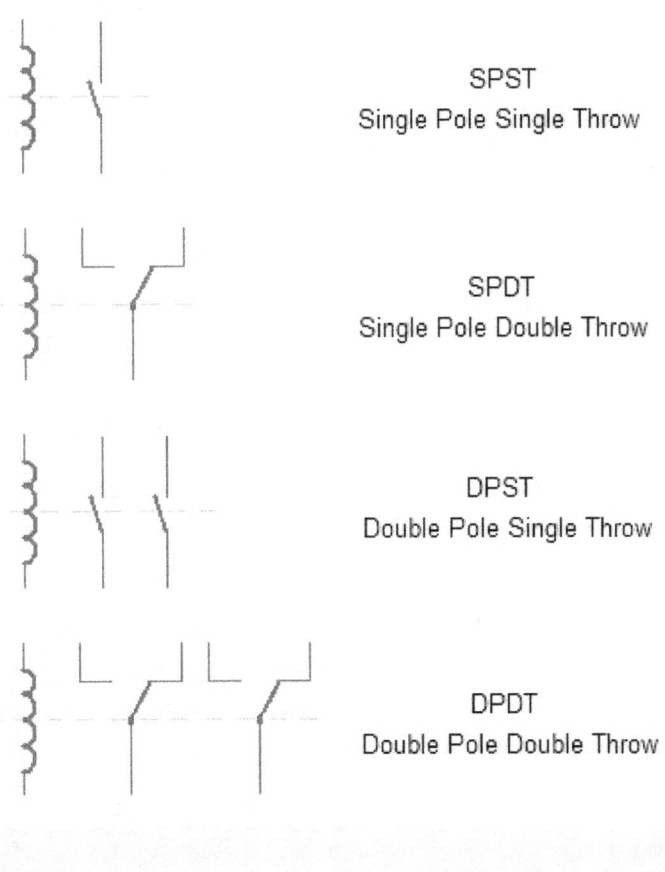

SPST
Single Pole Single Throw

SPDT
Single Pole Double Throw

DPST
Double Pole Single Throw

DPDT
Double Pole Double Throw

Single Throw relays are designated either as Normally Closed (NC) or Normally Opened (NO). The coil in a Normally Closed relay must be energized to open the relay, and the coil in a Normally Open relay must be energized to close the relay. Double Throw relays will have terminals labeled as Normally Opened and Normally Closed as well as a third terminal labeled Common. A Double Throw relay, therefore, can be used as either a Normally Closed or Normally Opened Single Throw relay simply by leaving one of the terminals unconnected.

One additional type of relay is Latching. Unlike traditional relays, latching relays maintain their current state, regardless if power is applied. A latching relay works either with two separate coils which are momentarily energized to independently close or open the relay or with a single coil which is momentarily energized to change the current state of the relay. One advantage of latching relays is that they are more power efficient as power is only required to change states (as opposed to maintain a state). One drawback is that more circuitry or logic may be required to monitor which state the relay is currently in. Latching relays are also poor candidates for safety critical applications where typical relays may be used to break open a line if power is lost to a sub-system as a "fail safe" condition.

HW_Q5:
Describe a transistor in one word.

HW_A5:
Switch or Amplifier

Any 'one word' type questions should generally be answered as such. This is one situation where it is better to hold back on an elaborate answer to demonstrate your knowledge or, if you want to go into more depth, do so only after you have given your one word answer. If you immediately go into a long winded response detailing the semiconductor process involved in manufacturing an NPN transistor or start discussing the implications of electron tunneling, the interviewer may start to question if your knowledge is purely academic, or if you actually know how to use one of these devices.

These types of questions are used to get a feel for your practical understanding of electronics. You've graduated with an EE degree, so you clearly have some book smarts and were able to pass many math and theory related tests, but do you know what these parts actually do? A surprising number of fresh graduates can talk extensively about transistors without being able to adequately describe how to use one to switch on an LED.

HW_Q6:

Can you name the three terminals of the transistor below?

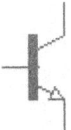

HW_A6:

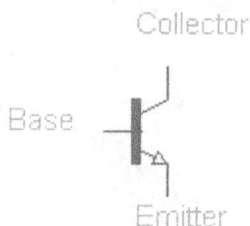

Collector

Base

Emitter

NPN Transistor

Before the interview, review and memorize (if it not already burned into your head) the three terminals for both NPN and PNP bipolar junction transistors. Also know the three terminals for both N-Channel and P-Channel FETs (Field-Effect Transistors: gate, source, drain).

You may be able to get away with claiming "I'd look that fact up in a real world design situation" for some more obscure pieces of information, but transistor use is so common in electronic circuit design that you will be expected to know this.

Some engineers seem to take a preference to either BJT or MOSFET transistors, and you may see questions with significantly more of one or the other, so make sure you are familiar with both. You should also be familiar with the differences between the two and in what types of applications one may have an advantage over another.

BJTs are current driven (the base current and the transistor's gain will determine the collector to emitter current) and so can be imagined as a current amplifier. MOSFETs are voltage driven, and can be imagined as a variable resistor that is controlled by the gate to source voltage.

The power dissipated by a BJT is dependent on the collector to emitter voltage drop of the transistor, while the power dissipated by a MOSFET is dependent on its on-resistance. Newer MOSFETs can have extremely low on-resistances, making them more efficient, particularly for lower current applications. Their efficiency advantage really shines, however, in high frequency applications, as their switching speeds can be order of magnitudes faster than that of BJTs.

BJTs tend to be more robust than MOSFETs in some applications, although modern MOSFETs are being made more and more robust. BJTs typically require a much lower voltage to fully engage (0.7V) than MOSFETs (typically ~3V+) and so may be better suited for driving with low voltage sources, such as a microcontroller output. High current BJT applications may require a significant amount of base drive current. Although only a miniscule amount of current flows into a MOSFET's gate pin in a steady state condition, the gate can be capacitive, particularly for larger (higher power) MOSFETs, and so may require special gate drive circuitry to provide the initial surge of current needed to charge up the gate quickly. The voltage drop across BJTs lowers with increasing temperature. You should not use BJTs in parallel, therefore, because increased current flow through one transistor will cause it to heat up resulting in ever more current flow (until its magic smoke is released). Thermal runaway may lead to the destruction of the BJT. The resistance of MOSFETs, on the other hand, increases with temperature, which allows them to be used in parallel more easily.

Another note is that some types of MOSFETs, such as most Trench type MOSFETs, are designed specifically for use in applications where they are switched fully on or off, and using them as an analog amplifier can result in their destruction (see the *Safe Operating Area* in a MOSFETs datasheet to ensure it will not operate outside this area for a given application).

Also be aware of a major difference between JFETs (Normally On) and MOSFETs (Normally Off), though you will rarely see questions about JFETs.

While many students will be exposed to the physical construction of transistors, this is rarely brought up in interviews, unless this interview is specifically for a semiconductor design position.

NPNPNPNPNP...

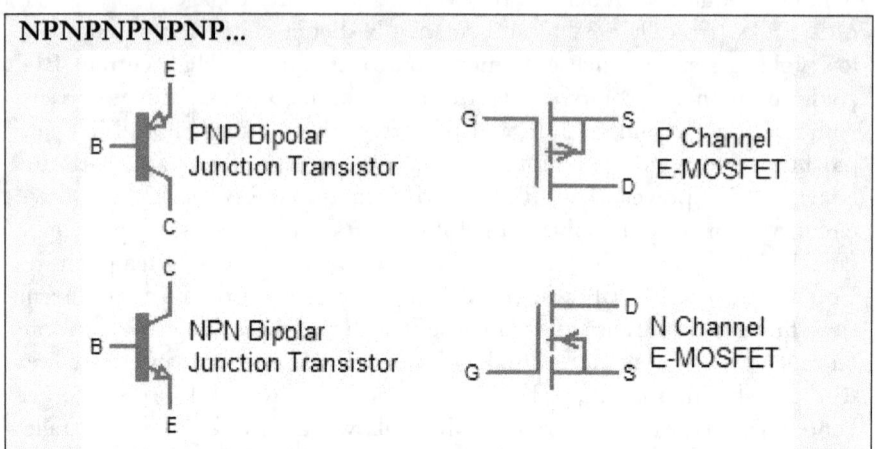

I, like seemingly many other engineers, have poor rote memory skills. Regardless of the depth of your transistor knowledge when it comes to their physical operation, however, you simply will not sound professional or knowledgeable when talking about how the current is flowing out of the thingamajig into the whatchamacallit and causing the voltage at the doohickey to be... (even if your physical and mathematical analysis is right). So make sure to have the proper transistor terminal names memorized. In typical electrical engineering fashion, the arrow orientation of transistors and FETs does not seem to be consistent, so pay special attention to the NPN / PNP and N-Channel / P-Channel arrow direction.

HW_Q7:

Assume that in the below circuit, pressing switch S7 is supposed to turn on the string of LEDs.

What is wrong with this circuit? (Assume a 2V drop across each LED)

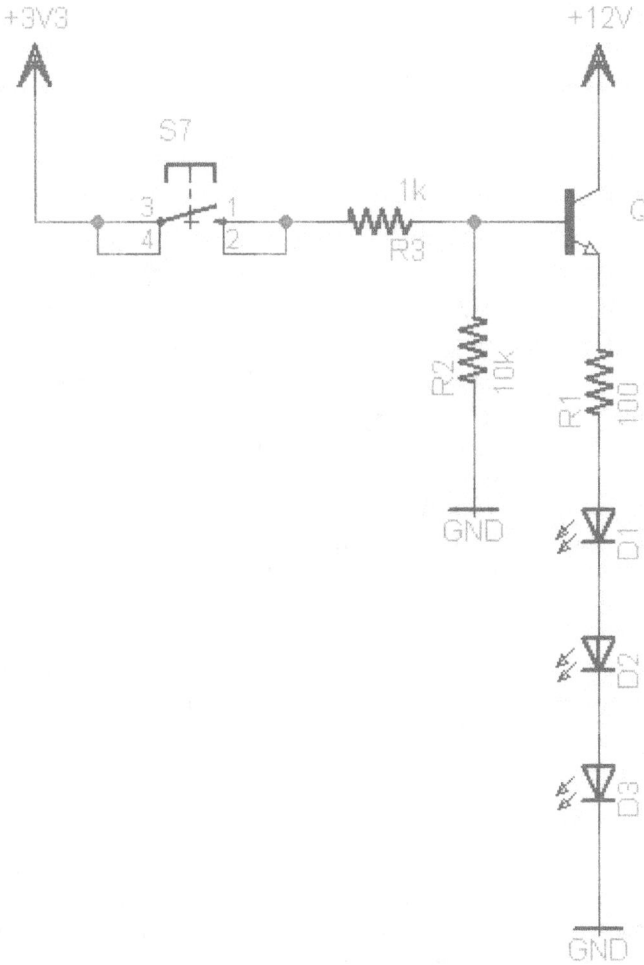

HW_A7:

This circuit will not work as designed because the voltage at the transistor's emitter (even when the transistor is turned on) will never be high enough to overcome the LED forward voltage (3 x 2V = 6V) due to the fact that the emitter voltage is always a minimum of approximately one diode drop (~0.7V) below the base voltage of a transistor.

Three possible solutions are:
1) move the transistor to the low side of the LEDs
2) switch on the transistor with the 12V rail, or
3) redesign the circuit to use a PNP transistor, with the switch pulling the base to GND.

This circuit tests you understanding of transistors, specifically the difference between the use of NPN and PNP transistors. The circuit is drawn here with a switch, but may alternatively be described as receiving output from a 3.3V microcontroller output pin. In fact, the circuit as drawn does not seem to make much sense. Since there is not a significant amount of current driving the LEDs, why not just use the switch to directly switch on the LEDs and get rid of the transistor altogether?

If the thought enters your mind you should say it. Your interviewers will most likely laugh and quickly explain it is an imaginary circuit and they are interested in your transistor knowledge, but will be impressed with your outside the box thinking.

If the problem with this circuit is not immediately obvious, walk through the circuit assuming a pressed switch. It is a good idea to jot down voltages on the schematic to demonstrate your understanding of the circuit and to help you visualize and work through the problem.

As seen in the following diagram, when the switch is pressed, the voltage at the base of the transistor will see a voltage divider with R3 and R2. The exact voltage at the base is not terribly important. We know from the voltage divider rule that the voltage will be 3.3V x 10kOhm / (10kOhm + 1kOhm) or approximately 90% x 3.3V or ~3V.

As we know, BJTs can be thought of as current amplifiers. The collector current will be a multiple of the base current as dictated by the gain of the transistor (the current gain is commonly specified as βF or hFE). For the base current to flow, however, there must be a voltage drop across the base to emitter junction. This is approximately a diode voltage drop or ~0.7V. This means that the voltage at the emitter in our circuit can only be maximum of approximately 2.3V, not nearly enough to overcome the 6V drop required to drive the LEDs (2V x 3 LEDs).

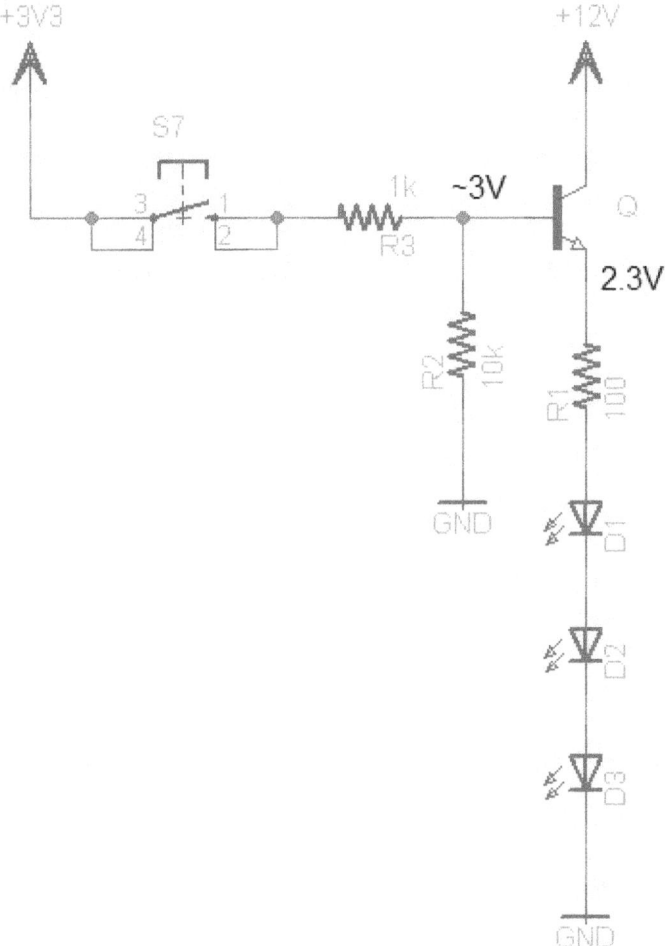

There are some different ways to solve this problem with a better transistor configuration. One is to move the transistor to the low side of the LEDs. As seen in the following diagram, this puts the emitter voltage at ground, and requires less than 1V to fully saturate the transistor and flow current through the LEDs.

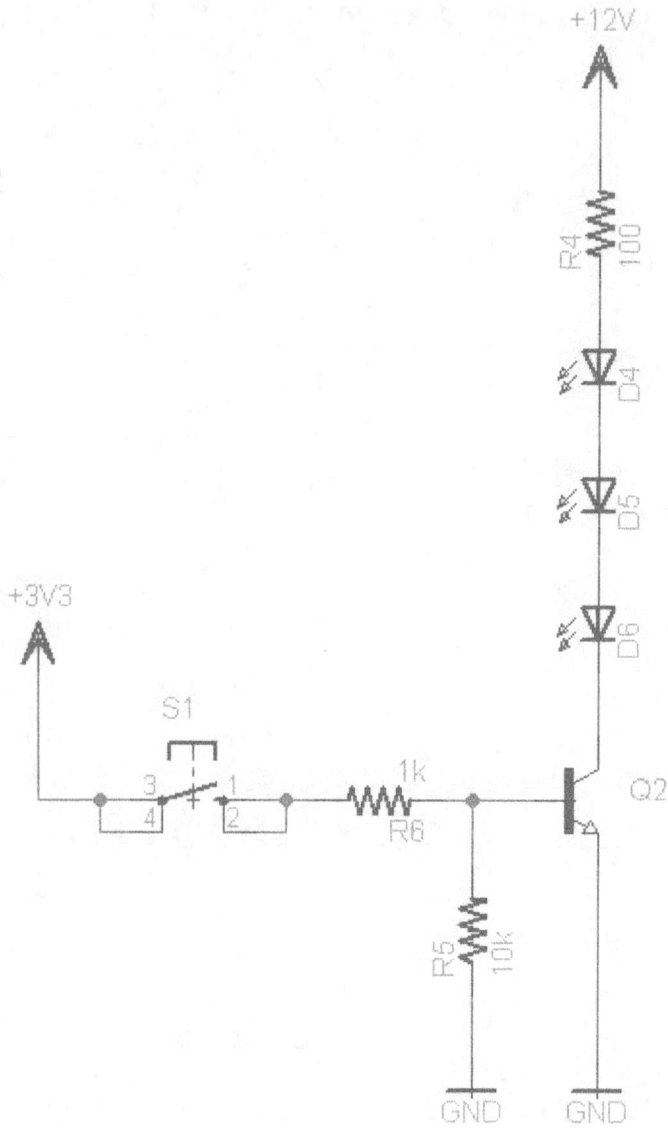

Note that the base voltage in this case will be around 0.7V (not 3V as calculated by the voltage divider above) because any voltage higher than that will cause current flow through the base to emitter junction of the transistor. In this case, R6 limits the transistor base current to (3.3V - 0.7V) / 1,000Ohms = 2.6mA. The gain of the transistor multiplied by this base current yields the collector current, although this transistor is likely easily saturated, and the current will instead be limited by R4 and the voltage drop across the string of LEDs.

Another solution is to replace the NPN transistor with a PNP transistor and change the base driving circuitry as displayed below.

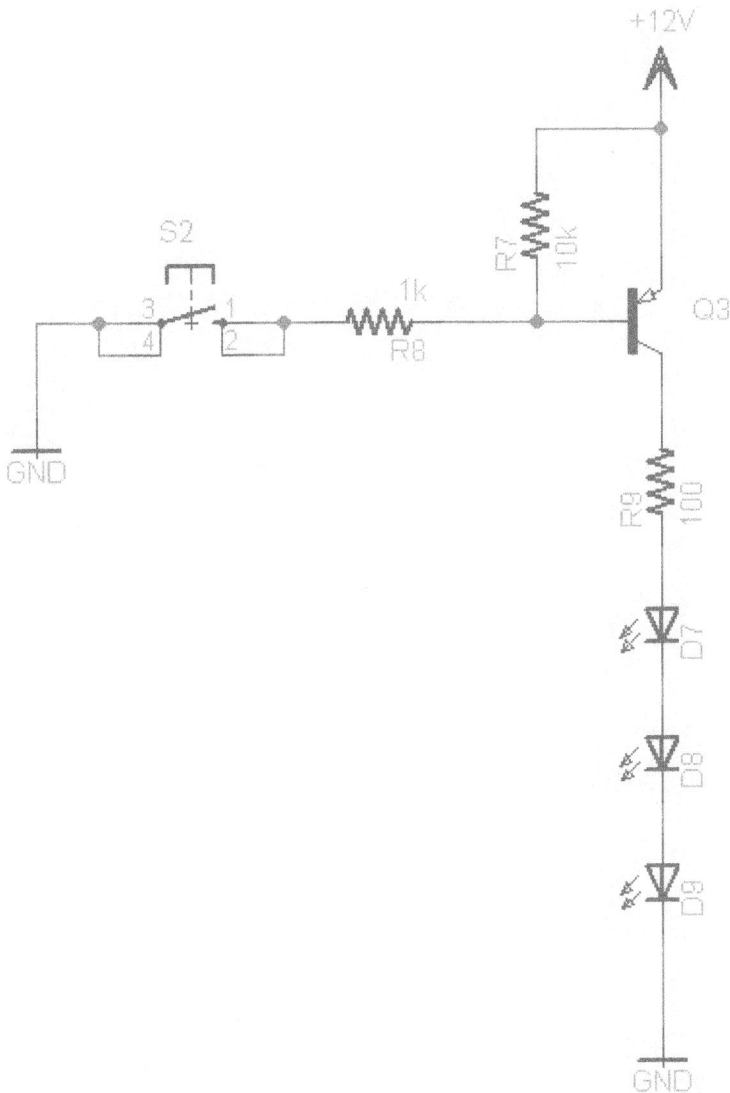

In this circuit, the base of the PNP transistor must be pulled approximately 0.7V below the emitter voltage to turn it on, which will be accomplished by pressing S2.

HW_Q8:

Given the corrected LED circuit from the previous question, what amount of current will flow through the LEDs when the transistor is turned on?

HW_A8:
60mA

The trick here is to remember to take into account the voltage drop across the LEDs (something that many seem to forget) 12V - (2V x 3) = 6V. Remembering Ohm's law, V = IR, and so I = V/R = 6V/100Ohms = 0.06A or 60mA.

If you really want to impress your interviewers, you can also take into account the collector to emitter voltage drop of the transistor (typically around 0.3V). This would give a total voltage drop of 6.3V, and a resulting current of (12V - 6.3V)/100 Ohms = 57mA.

HW_Q9:

If the input impedance of a MOSFET gate is so high, why are dedicated, high current gate driver components required in many high speed switching applications?

HW_A9:

In steady state operation, MOSFET gate current is extremely small. MOSFET gates can be capacitive, however, and so may require large slugs of current to charge and discharge in high speed switching applications. This is especially true when the MOSEFT must be switched very fast.

This question may come up while reviewing and discussing a schematic, say of a motor driver or switching power supply.

MOSFET Gate Driving

The typical gate voltage during the charging of a MOSFET Gate is displayed below.

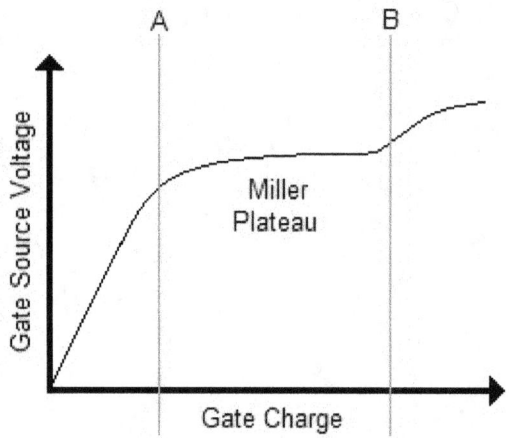

One interesting aspect of the charge curve is the Miller Plateau. The Miller Plateau results from the Miller Effect, which describes the increased input capacitance of an inverting amplifier. Once the gate voltage reaches the transistor's threshold voltage and starts to turn on the MOSFET, the Miller Effect comes into play and the increased input capacitance causes the Gate voltage to plateau until the MOSFET is completely turned on.

HW_Q10:

What is the difference between the active and saturation regions of a transistor?

HW_A10:

This can be confusing (and is sort of a trick question) because of the difference in terminology between BJTs and MOSFETs. I have not heard this specific question asked in an interview, but it does routinely come up in conversations about transistors, especially the term "saturated", and so it is important to have a proper understanding of correct terminology.

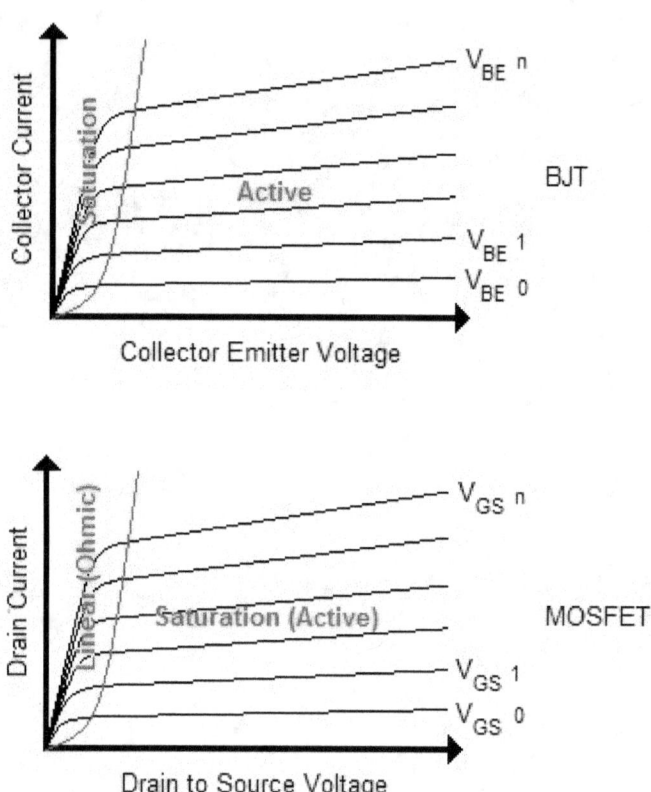

When a BJT is in its saturation region, the transistor is fully on. When it is fully off, it is in the cutoff region. When in its active region, the collector current is controlled by the base current (which is controlled by the base to emitter voltage). This state exists between the fully on and fully off states. BJTs are commonly said to be operating as an amplifier when in the active region, and operating as a switch when in the saturation region.

This terminology is somewhat intuitive, unlike for the MOSFET, which reverses the use of the word saturation. A MOSFET used as a switch will be used in the linear or triode region. This region is also called the ohmic region because the MOSFET acts as a variable resistor. In the saturation or active region, the MOSFET acts as a current source in which the current is controlled by gate to source voltage.

It is not uncommon to hear even experienced engineers refer to a fully on (or fully enhanced) MOSFET as being "saturated", referring to the "full extent" meaning of the English word, although this is confusing when considering its mode of operation.

HW_Q11:
What is the difference between resistance, reactance and impedance?

HW_A11:

Resistance is the opposition to the flow of electrons through a conductor. It is commonly referred to as the electrical equivalent of friction.

Reactance is the opposition to the change of current, and therefore is dependent on frequency. Unlike pure resistance, which only affects the amplitude of a signal, reactance affects both the amplitude and phase of a signal. Reactance is caused by capacitive and inductive circuit elements. The phase of the voltage across a capacitor lags current by 90°. The phase of the voltage across an inductor leads current by 90°. Reactance is commonly referred to as the electrical equivalent to inertia.

Impedance is the combination of both resistance and reactance, sometimes referred to as "resistance at frequency".

HW_Q12:
What is the circuit below?

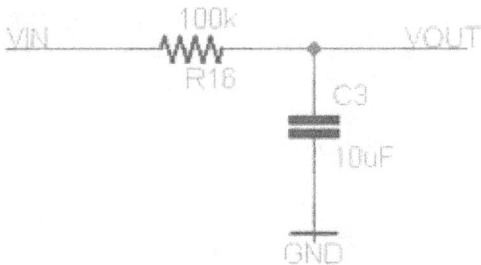

HW_A12:
An RC filter, or specifically, a low pass filter.

You should be familiar with common first order filters which can be designed with resistors, capacitors and inductors. Low pass filters are extremely common in real world designs, so some extra study into these is warranted.

It can help to think what will happen at frequency extremes. At high frequencies, the capacitor will look like a short circuit and shunt those frequencies to ground. At low frequency or DC, the capacitor will look like an open circuit, and therefor will have no effect on the input signal voltage.

HW_Q13:

What is the time constant and cutoff frequency of this filter? (Circuit from HW_Q12)

HW_A13:
Time constant: 1 second
Cutoff frequency: ~0.159 seconds

The time constant for an RC filter is simply R (in Ohms) x C (in Farads), which in this case works out to be 1 second. The cutoff frequency is:

$$fc = 1 / (2\pi \text{ x } R \text{ x } C)$$

This is the frequency, in hertz, that the signal will be attenuated by approximately 30% (-3dB).

These are two equations worth memorizing given how prevalent RC filters are in electronic designs.

How much attenuation?

It can be difficult to remember whether an RC filter attenuates a signal by 50% or 30% at the cutoff frequency. -3dB corresponds to 50% power attenuation, and 30% voltage attenuation. The power attenuation considers both voltage and current, but in dealing with filters we are generally concerned only with the voltage.

Speaking about filter attenuation (or amplifier amplification) among other things in decibels (dB) can be confusing, even for working engineers. Despite their prevalence in documentation and literature, it may just be easier to ignore them altogether, but having a good grasp of this concept will help impress your interviewers. A great, free, document to give you a practical understanding of decibels is:

Rohde & Schwarz Application Note *1MA98, dB or not dB? Everything you ever wanted to know about decibels but were afraid to ask...*

Freely available on the Internet. This document includes a table of common dB values with their corresponding power ratio and voltage ratio I like to keep pinned to my office wall.

Another great introduction to decibels is David Jones Electronic Engineering Video Blog episode 49: *Decibels (dB's) for Engineers – A Tutorial* (available at www.eevblog.com and on YouTube).

HW_Q14:

Assuming a variable resistance load at the end of a long power transmission cable with 50Ohms of resistance, what load resistance will allow you to achieve maximum power transfer to the load?

HW_A14:
50 Ohm

According to the maximum power transfer theorem, maximum power is transferred when the load resistance is equal to the source resistance.

You may be correct in thinking that a lower resistance load will allow for more current flow through the load, but the reduced voltage drop across the load results in less power.

HW_Q15:

Considering the AC/DC conversion circuit below, draw the voltage waveforms:

1) On the secondary side of the transformer
2) After the rectifying diode bridge assuming that C2 was not present
3) After the rectifying bridge assuming C2 is present

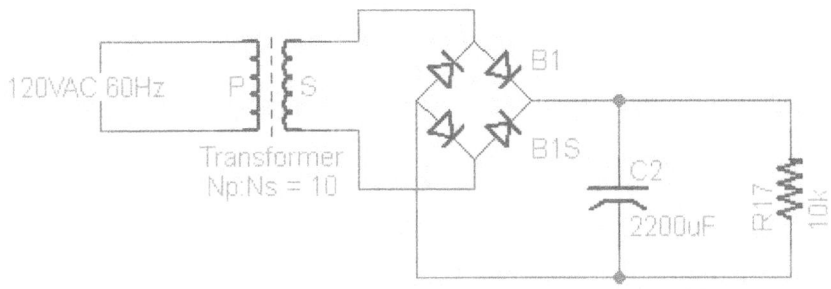

HW_A15:

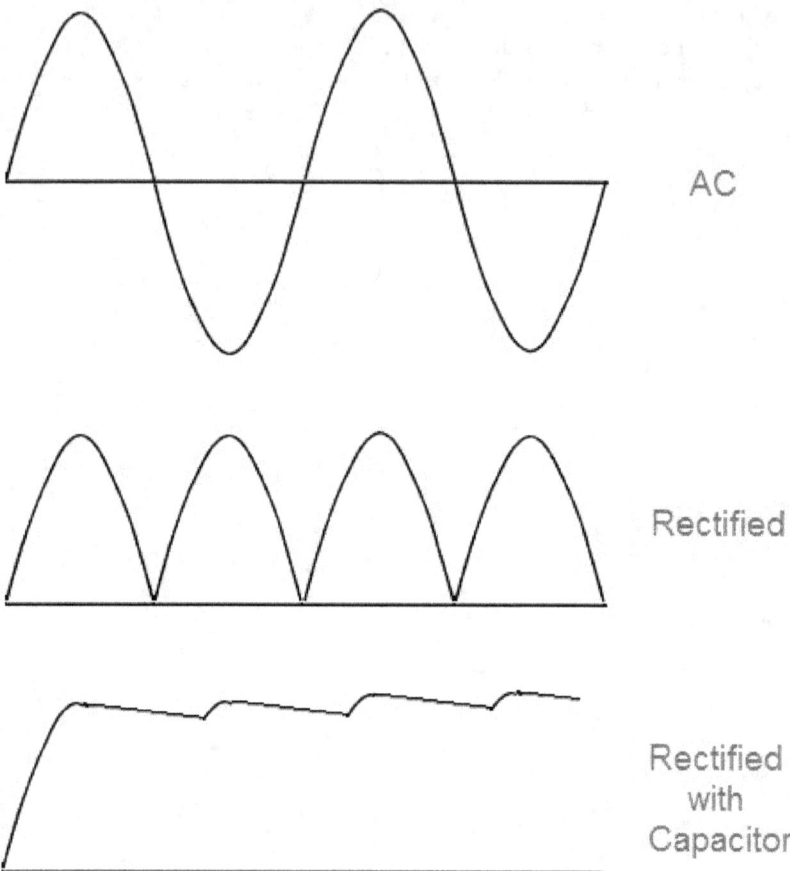

AC

Rectified

Rectified
with
Capacitor

This is a simple linear AC to DC converter. A transformer is used to step down the line voltage. The ratio of primary to secondary turns dictates the step down voltage. In this case, the primary to secondary ratio is 10:1, stepping down the 120VAC input to approximately 12VAC on the secondary side.

A diode bridge is used to rectify the AC waveform to a single sided waveform and the bulk capacitance of C2 helps smooth the output voltage. Note that because of the small voltage drop across the diodes, the rectified voltage will drop to around 0.7 (i.e. it will not reach exactly 0V).

Take careful note of the orientation of the diodes in the bridge. While most candidates are aware of the diode bridge functionality, many have trouble accurately reproducing the diode configuration from memory.

HW_Q16:

What is the equivalent DC voltage of a sine wave AC power source with peak voltages of +/- 170V?

HW_A16:
Approximately 120V DC.

In this case, the voltage was chosen to match the typical line (mains) voltage in the US, assuming that little to no math would be required provided you have understanding and knowledge of the US electrical distribution system and the difference between RMS (root mean square) and peak voltage readings. The RMS voltage of an AC power source corresponds to the DC voltage of an equivalent DC power supply. In other words, for a given load, you could provide an equal amount power with a +/-170V peak AC supply or a 120V DC supply. For a sine wave, the RMS voltage is approximately $1 / \sqrt{2}$ or 0.707 times the peak voltage. Remembering that for a sine wave, the RMS voltage is roughly 70% the peak voltage, is usually sufficient.

This question could lead into a discussion about multimeters and their use for measuring AC voltage. Most multimeters will display AC voltage in RMS format. Many, however, calculate the RMS voltage assuming a perfect sine wave. If you are measuring an AC voltage with a wave form that varies from a perfect sine wave, a corresponding error will be present in the reading. Some multimeters are marketed as "True RMS" meters, and while generally more expensive, are able to calculate more accurate RMS values on non-perfect sine waveforms.

HW_Q17:
What is the difference between Watts and Volt-Amps?

HW_A17:

Watts is a measure of real power (the power that can be used by the load to perform work).

Volt-Amps is a measure of the apparent power (voltage times current).

For a purely resistive load, these two values will be identical. A highly inductive load, on the other hand, such as a motor or large power supply, can cause some phase delay which results in excess current on the AC power feed. In this case, a current reading on the power feed will show a larger current than is actually used by the load, because some current is simply being pushed back and forth (inductors resist current change). This is wasteful, as it increases resistive loses and puts additional burden on the power distribution system.

HW_Q18:
What is the purpose of PFC (Power Factor Correction)?

HW_A18:

Power Factor is the ratio of real power to apparent power, typically given as a fraction between 0 and 1.

An electrical system supplying many loads with poor power factors experiences higher current demands and may result in reduced performance and/or efficiency of the power supplies sharing the system. Large capacitor banks may be required to deal with the poor power factor, and it is not uncommon for electrical distribution companies to charge extra fees to customers with low power factor loads. To prevent the need for overbuilt electrical distribution systems and added distribution costs, regulations have been put into place for many common commercial and industrial products to ensure high power factors.

This is accomplished with power factor correction. The purpose of which is to make a load appear to be purely resistive, so that the load's real (or true) and apparent power are equal, and the power factor ratio is close to 1. Passive power factor correction may accomplish this with a filter, while Active power factor correction uses sophisticated circuitry to keep the supply current in phase with the supply voltage by controlling the duty cycle of a high frequency converter.

If you are interviewing for a position that involves power supply design, this is a subject that will be worth some additional research.

HW_Q19:

Match the power supply switching topologies shown below to:
Buck
Buck-boost
Boost

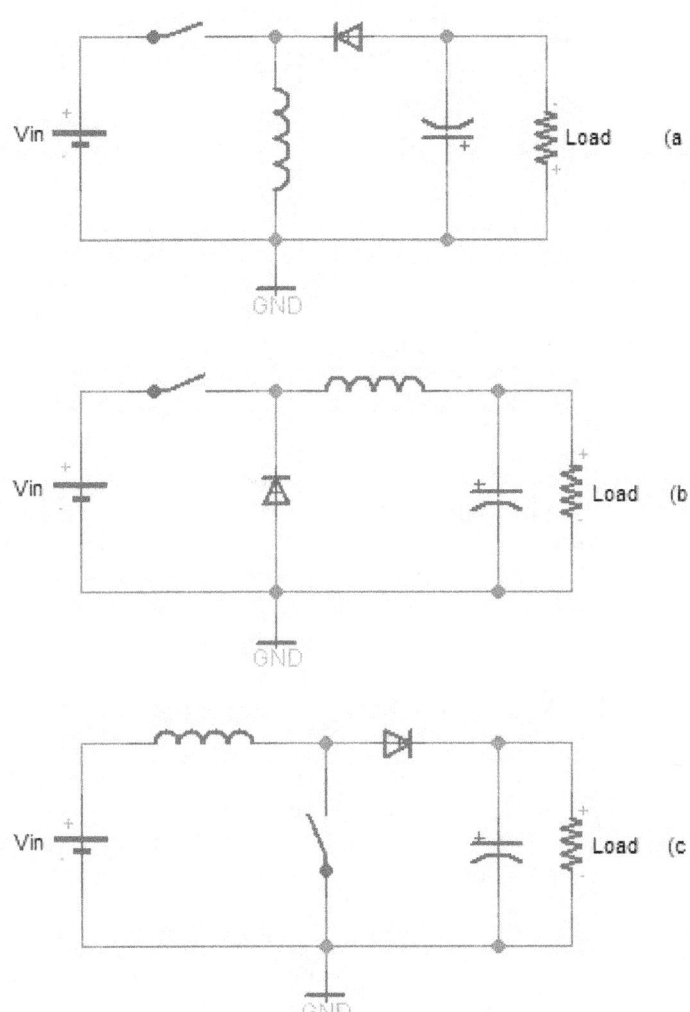

HW_A19:
> Buck: b
> Buck-boost: a
> Boost: c

While there are other power supply topologies, the Buck, Boost, and Buck-boost are the three most popular, with the Buck and Boost being the most likely to see/discuss in an interview.

Note that in the simplified diagrams above, the switching element is represented with a simple switch symbol, but that in a real schematic this would be implemented with a MOSFET or similar solid state transistor technology.

Two additional power topologies worth mentioning are the Flyback converter and the SEPIC converter. The Flyback converter is really just a buck-boost converter that uses a transformer in place of an inductor to provide isolation between the input and output. The SEPIC (Single Ended Primary Inductor Converter) is a unique and somewhat complex converter that can generate an output either less than or greater than its input voltage, without reversing polarity. Historically, this converter has not been used much due to the relatively complex control analysis required to prove stability with dynamic loads. It has become more popular in recent history, however, with the ever increasing use of low voltage battery powered products. A SEPIC converter can be used, for example, to provide a 3.3V rail off of a single cell battery with a voltage that may start at 4.2V at full charge and drop to below 3V before it is considered empty.

REGUALTORS, Mount Up! - Linear vs Switching

SMPS (Switch Mode Power Supplies) are becoming ever more popular in modern, power conscious designs. Their increased adoption has led to companies selling power supply control ICs (such as Linear Technology and Texas Instruments, among others) to offer ever more capable and easy to use products.

While in the past, designing a SMPS required significant power supply and control expertise, the new products available greatly simply the process or even allow for completely integrated and modular solutions that can be dropped into a design.

That being said, there is still a place for linear regulators. When only needing to supply a small amount of power, linear regulators offer a simple,

---

low cost and small solution. Linear regulators may also have an advantage in applications that require extremely low noise or fast response.

In applications with higher power requirements, however, heat sinking demands can drive up the cost and size of linear solutions. In particular for higher power applications, switching regulators offer higher efficiency, smaller size and typically lower cost than an equivalent linear solution.

HW_Q20:
For each of the three previous power supply topologies (HW_Q19), draw the current flow both with the switch open and closed.

HW_A20:

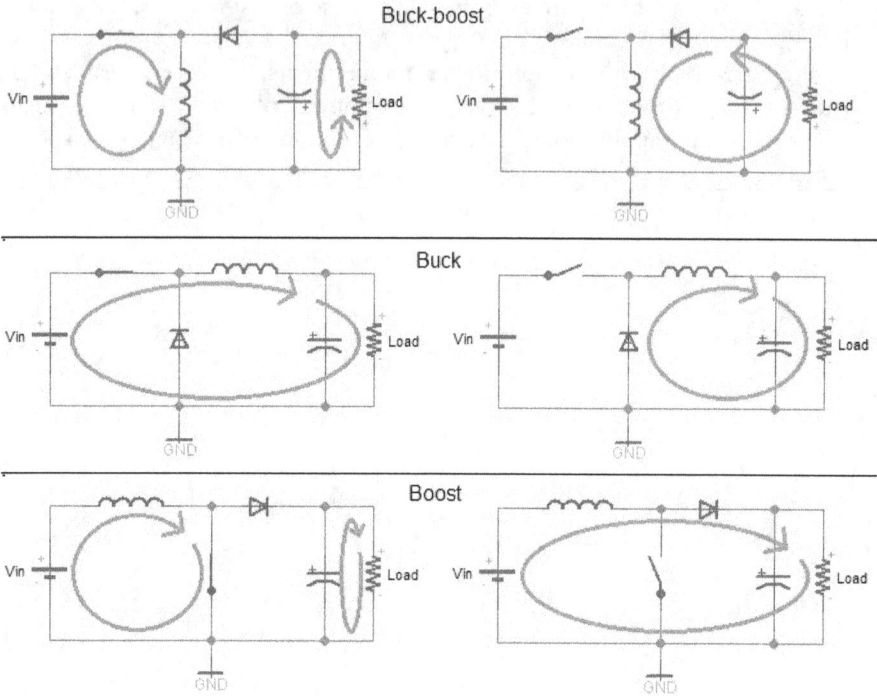

Current flow for switching power supplies is important to understand for PCB layout, where small current loops are needed to prevent a product from failing radiated emissions regulatory testing.

The most critical pathways to be concerned about are those where current is switching. In the Buck converter, for example, current is always flowing through the inductor, and so its placement on the PCB is not as critical as that of the diode, which should be placed closed to the switch (usually a MOSFET) to minimize the switched current loop.

Thinking about current flow in these converters will help drive a more intuitive understanding of their operation.

HW_Q21:

What is the name of the below circuit? What is it used for?

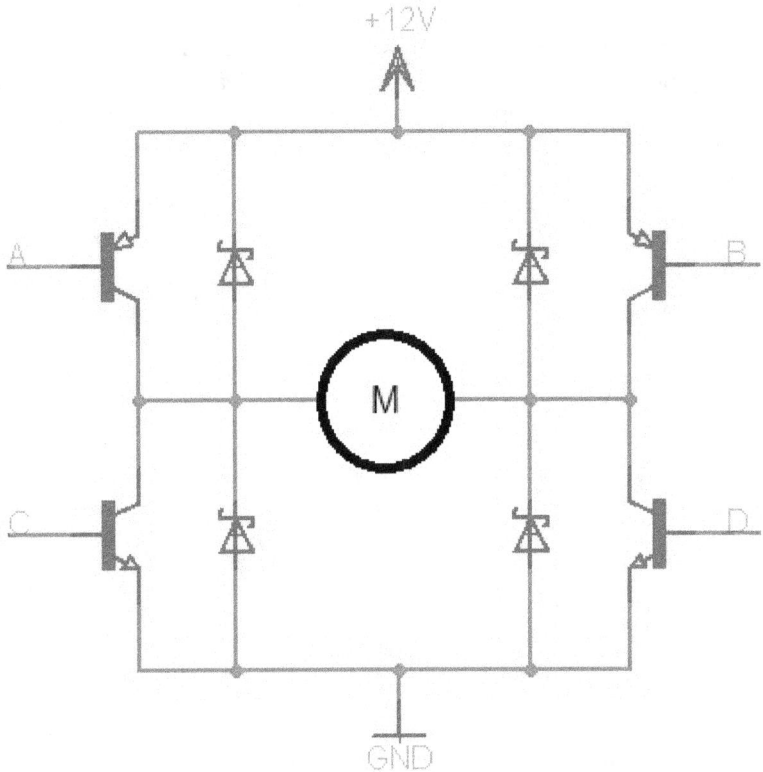

HW_A21:

An H-Bridge. This is used to drive a DC motor or similar load in which it is desirable to be able to flow current in either direction through the load.

Turning on transistors A and D, for example, will spin a DC motor in the opposite direction as when turning on transistors B and C. Turning on either A and C or B and D simultaneously should be avoided in this type of circuit as it will cause a power supply short circuit, possibly damaging the transistors, supply or both.

HW_Q22:

What is the purpose of the Schottky diodes in the previous H-Bridge circuit (HW_Q21)?

HW_A22:

The diodes help prevent voltage spikes from damaging the transistors.

The DC motor is inductive. Inductors resist changes in current. Therefore, when the transistors are quickly turned off a large voltage spike will be created across the motor. The diodes provide a path for the current to flow and prevent a large and potentially destructive voltage from forming.

HW_Q23:

Assuming VIN = 2V, what is VOUT in the op amp circuit below?

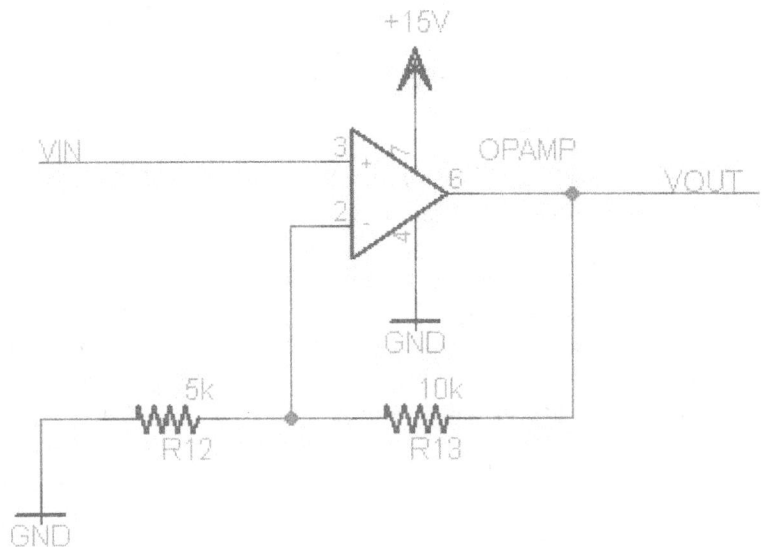

HW_A23:
6V

This op amp is configured as a non-inverting amplifier. Two op amp configurations you should be familiar with by sight are the inverting amplifier and non-inverting amplifier configurations. These are very commonly used circuits.

The output of the non-inverting amplifier is Vin x (Rlow + Rhigh) / Rlow (where for this circuit, Rlow = R12 and Rhigh = R13).

You don't have to memorize this equation, however, if you remember basic op amp principles. After seeing this op amp is configured in negative feedback mode, you know the op amp will attempt to keep the negative input equal to the positive (V- = V+). Therefore, if the positive input is set to 2V by VIN, the op amp will adjust its output to make the negative input 2V as well.

The voltage divider rule allows us to solve for V- as a function of Vout:
V- = Vout (Rlow / (Rlow + Rhigh))

But what we really want to know is Vout as a function of Vin. Remembering that in negative feedback mode V- = V+, and that in this non-inverting amplifier circuit V+ = Vin, we can substitute V- for Vin.
Vin = Vout (Rlow / (Rlow + Rhigh))
We can then use some basic algebra to solve for Vout as a function of Vin.
Vout = ((Rlow + Rhigh) / Rlow) Vin
This is occasionally further reduced in some textbooks to:
Vout = ((1 + Rhigh / Rlow) Vin

For the above circuit, therefore:
Vout = ((1 + 10kOhms / 5kOhms) 2V = 6V

Operational Amplifier Sources

A great source for common op amp circuits is National Instruments (since bought by Texas Instruments) *Application Note 31 - Op Amp Circuit Collection.*

A quick Google search will result in a PDF version of this app note, which should be high on your pre-interview study list.

Additional free application and reference notes that (while significantly longer) are great resources for learning about the practical use of op amps:

Texas Instruments *Op Amps for Everyone*, Ron Mancini.

Texas Instruments *Understanding Operational Amplifier Specifications*, Jim Karki.

Texas Instruments *Handbook of Operational Amplifier Applications*, Bruce Carter and Thomas R. Brown.

All freely available on the Internet.

HW_Q24:
Describe (in writing) X in terms of A, B, C and D.

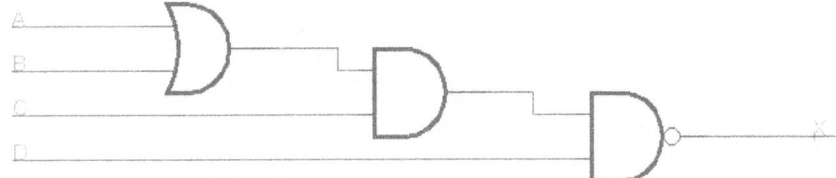

HW_A24:
X is NOT (D AND (C AND (A OR B)))

A logic question like this is generally used to see if you have a basic familiarity with logic components, and may help lead into a conversation about the difference between synchronous and asynchronous circuits.

Most modern designs do not use much 'jelly bean' logic in terms of dedicated ICs (at least anywhere close to the extent that they were used in the past, before microcontrollers became so prevalent), but an occasional logic chip may make for an elegant solution to certain problems.

In addition, a solid understanding of digital logic fundamentals is still in high demand for work on or with programmable logic devices such as CPLDs (Complex Programmable Logic Devices) or FPGAs (Field-Programmable Gate Arrays). Some software packages allow graphically designing logic for such devices using standard logic elements, but more complex applications are programmed with a hardware description language such as Verilog or VHDL.

HW_Q25:

A given linear voltage regulator has a $R_{\theta JA}$ (junction-to-ambient thermal resistance) of 51.5°C/W. What is the maximum power handling capability of this part assuming a maximum junction temperature of 125°C and a maximum ambient temperature of 60°C.

HW_A25:
1.26 W

(Junction – Ambient) / θJA = (125°C – 60°C) / 51.5°C/W = 1.26W

These types of calculations are common in real world electronics design, where thermal dissipation is a common issue. As previously stated, it is worth mentioning that this calculation leaves no extra headroom and that in a real world design other factors, such as heat given off by nearby components, airflow within the assembly, etc., would have to be considered.

HW_Q26:

Given an analog input signal with a maximum frequency of 8kHz, what is the minimum ADC (Analog to Digital Converter) sampling frequency required to prevent information loss?

HW_A26:

16kHz (with a big asterisk).

This is always a popular question and in many cases, it seems the interviewer just wants to hear the name "Nyquist" uttered by the interviewee. Another version of this question may be:

Given an ADC with a 16kHz sample rate, what is the maximum cutoff frequency its low-pass input filter must be designed to?

Being able to expound on this answer, however, will give you a big leg up on many other recent graduates (and even some experienced engineers).

It is common DSP (Digital Signal Processing) knowledge that when digitizing an analog signal, we must sample the signal at a minimum of two times its frequency or risk losing information. Along the same lines, we must low-pass filter an analog signal to at least one half the sampling frequency of an ADC or risk observing bogus information due to aliasing (if this is not intuitive to you, there are some good YouTube videos on the subject that animate this effect nicely – search for "nyquist frequency and aliasing").

To digitize an analog signal with a bandwidth of 8kHz, therefore, we must sample at a minimum of 16kHz. Or given an ADC with a 16kHz sample rate, we must place an 8kHz low-pass filter in front of the ADC.

Most textbooks discussing this subject do so assuming imaginary, perfect filters that are able to completely block frequencies above their cutoff frequency. In the real world, low-pass filters are specified with a -3dB cutoff frequency, at which point they attenuate the input signal by approximately 30% (meaning the majority of the signal is still getting through at that frequency!)

Beyond the cutoff frequency, the amplitude of the signal passing through the filter diminishes as a function of the order of the filter. To achieve hard cutoff, a high order filter is needed. This typically means a large, expensive circuit with many op amps and passive components.

In the past, it was not uncommon to see up to an eighth order low pass filter placed in front of an ADC to prevent aliasing, but as the performance of ADCs has improved and the price has decreased, it is now more common to see a first or second order low pass filter paired with a fast sampling ADC, with additional filtering occurring in software.

Free Sampling!

A great (and must) read paper regarding ADC interfacing is:
Sampling: What Nyquist Didn't Say, and What to Do About It
by Tim Wescott and available freely from his website at www.wescottdesign.com.

For those with an interest in Digital Signal Processing, there are also a couple of in depth and thorough resources available online:

The Scientist and Engineer's Guide to Digital Signal Processing, by Steven W. Smith, Ph.D. and freely available at www.dspguide.com.

Mixed-Signal and DSP Design Techniques, edited by Walt Kester and freely available on Analog Devices website (www.analog.com).

HW_Q27:
Given a 10-bit SAR ADC with the interface shown below, what is the minimum sample time required to achieve a conversion accurate to within 0.5 bits?

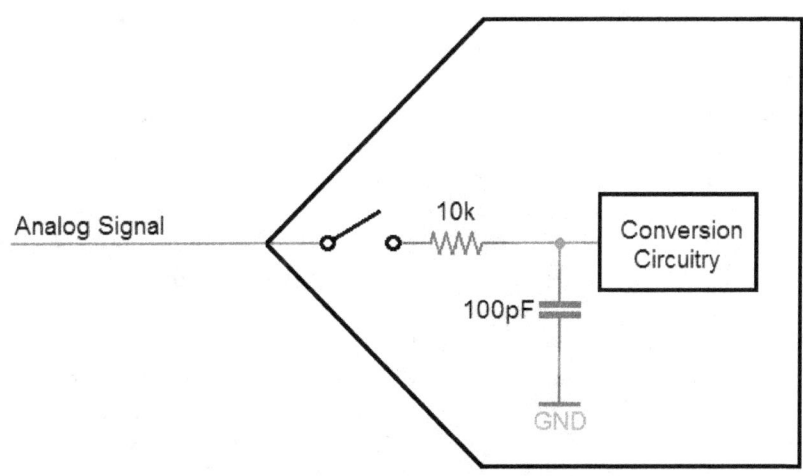

HW_A27:

This question is a personal favorite because so much of modern electrical engineering revolves around the boundary between our analog world and its digital representation, and because it provides another opportunity to discuss the RC filter! Even if you cannot get to the numerical answer, knowing the physics behind the problem can go a long way to impressing your interviewer.

The two most popular types of ADCs are SAR (Successive Approximation) and Sigma-Delta. Having a high level understanding of how these two types of ADCs work is worthwhile (and there are many explanations available on the Internet). Of the two, SAR ADCs are more popular for general purpose analog-to-digital conversion and more typically integrated into microcontrollers. One important aspect of their operation to be aware of is that they use a sampling capacitor to hold the signal voltage while conversion to a digital value takes place.

The input to the ADC will have some impedance, and this is especially true for ADCs that have a built in multiplexer, such as is common for those integrated into most microcontrollers. A microcontroller, for example, may only have a single ADC, but have multiple pins tied to it through a multiplexer. The input resistance is typically specified in the ADC's datasheet, and occasionally depicted in an analog input model similar to this question's figure.

The internal resistance and capacitance of the ADC create a low-pass RC filter. The sample capacitor must be charged up to the signal voltage prior to starting a conversion, or the conversion will not be an accurate representation of the signal voltage. For this reason, it is important to consider the sampling time (sometimes referred to as settling time) required for an application. This is typically a user configurable setting. In order to benefit from the maximum ADC resolution, the signal must settle to within less than 1 bit of resolution of the ADC.

You should know that the number of discrete values (the resolution) of an ADC is given by 2 to the power of the number of bits of that ADC. A 10-bit ADC, therefore, has a resolution of 2^{10} or 1,024 discrete values. Common ADC language can be a bit confusing when talking about error, which is typically communicated in bits representing the smallest discrete value of the ADC. For example, if a 10-bit ADC has a maximum specified offset error of 2 bits, it has up to $2/2^{10}$ or 2/1,024(full scale) offset error. It does not have up to $2^2/2^{10}$ or 4/1,024(full scale) offset error, which could mistakenly be assumed given how the resolution of ADCs is advertised.

For this particular question, we are trying to determine the sampling time required to reach 0.5 bits of resolution, which is 0.5/1,024 for this 10-bit converter.

The filtered voltage while charging as a function of time is:

$$V(t) = V_0(1\text{-}e^{\text{-}t/\tau})$$

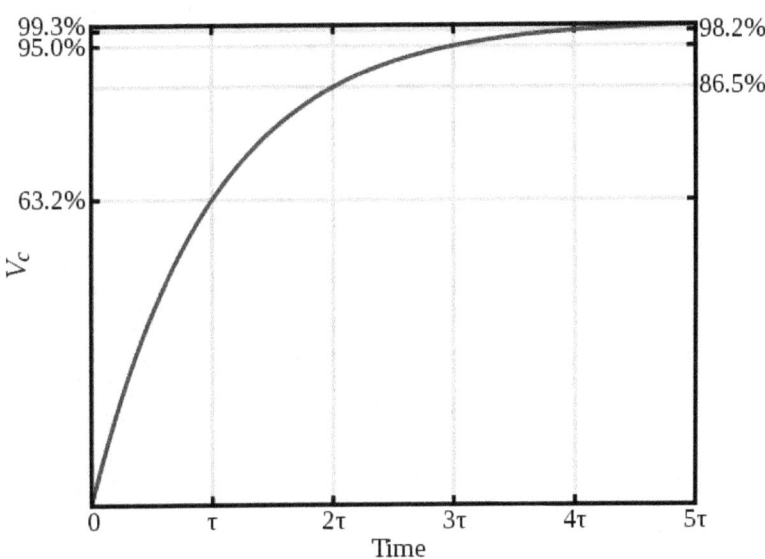

For completeness, the voltage while discharging as a function of time is:

$$V(t) = V_0(e^{-t/\tau})$$

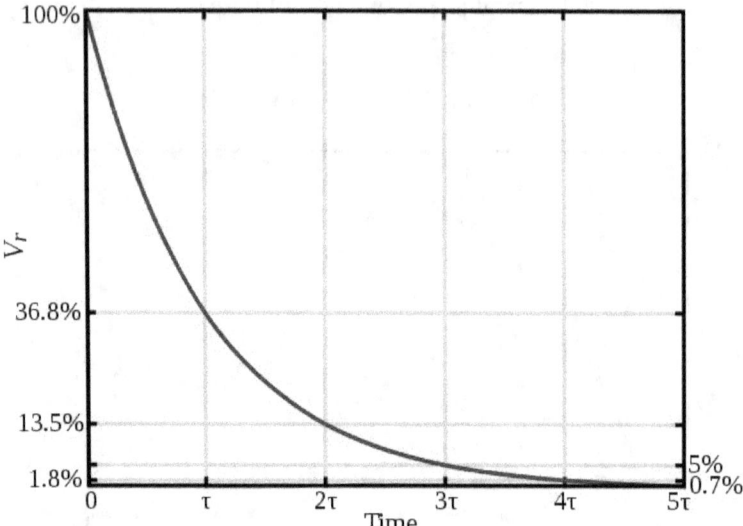

RC filters are so prevalent in electrical designs, that memorizing these two equations is worthwhile, although I wouldn't hold it against an interviewee for not having them memorized. These are definitely the type of equations you could easily look up with access to a computer. It is more important to have an intuitive understanding for what the charge and discharge curves of a RC filter look like.

Using the above formula, we can calculate the number of time constants needed to charge a 10-bit ADC up to within 0.5 bits of resolution (1023.5/1024):

1023.5/1024 = 1-e^{-t}
1023.5/1024 -1 = -e^{-t}
1 - 1023.5/1024 = e^{-t}
1/(1 - 1023.5/1024) = e^{t}
ln(1/(1 - 1023.5/1024)) = ~7.6 time constants

That is, it takes approximately 7.6 time constants to charge an RC filter up to within 1023.5/1024 of the input voltage. The time constant of an RC filter, τ (tau), is simply the resistance in Ohms times the capacitance in Farads. The time, therefore is:

(10k Ohms)(100pF)(7.6) = 7.6 us.

One important note is that this math assumes the signal to be measured comes from a low impedance source. If, on the other hand, the signal source has some resistance, that resistance would need to be added to the ADC's internal resistance to determine the appropriate sample time.

ADC References

A couple of good ADC application notes are mentioned below (freely available online).

Texas Instruments: Switched-Capacitor ADC Analog Input Calculations
This app note goes over math similar to that used to answer this question.

Atmel AVR127: Understanding ADC Parameters
This provides a good review of the many specifications that appear in a typical ADC datasheet, some of which are not patently obvious without some background knowledge. Having a solid understanding of these parameters will come in useful when trying to select an appropriate ADC for a real world application.

HW_Q28:
Explain what the P, I and D terms of a PID controller are and how these individual terms contribute to the control output.

HW_A28:

PID control, or a subset (such as PI), is commonly used in electrical engineering. In fact, the fast majority of electronic controllers in existence are using some form of PID control. Having a general idea of how the contribution of each term effects the output of a controller, and being able communicate it in an intuitive manner will give you a big leg up on most candidates. I find it easiest to think of the contribution of each term in terms of time.

P - Proportional
The proportional term provides response to the instantaneous error. The contribution of the proportional term will be the error (difference between the set-point and the measured value of whatever is under control in a particular system) multiplied by the proportional gain.

I – Integral
Integrating the error over time provides a view backwards in time. The contribution of the integral term will be driven both by the current and past error of the system. This helps to provide a long term control offset. The magnitude of the integral term's contribution can be controlled with the integral gain. Increasing the Integral term too much can result in sluggish response to error and overshooting the set point value.

D – Derivative
The derivative of the error is the rate of change of the error, which helps provide a view to the future state of the system. The derivative component of the controller typically works against the proportional and integral terms, helping to prevent overshooting the set-point when large steps in error cause the proportional and integral terms to grow large. The magnitude of the derivative term's contribution can be controlled with the derivative gain. The derivate term is especially susceptible to noise, and can cause instability if set too high in a noisy system.

It is important to note that while in many control textbooks all three terms operate on the system error, it is common for the derivative term in real world controllers to operate directly on the measured value.

PID *without a PhD* by Tim Wescott is a classic engineering article and great overview of both PID control and software implementation. The article is readily available on the Internet and is a must read for any engineer who may be involved with PID control.

HW_Q29:

What is the name of the type of plot displayed below?

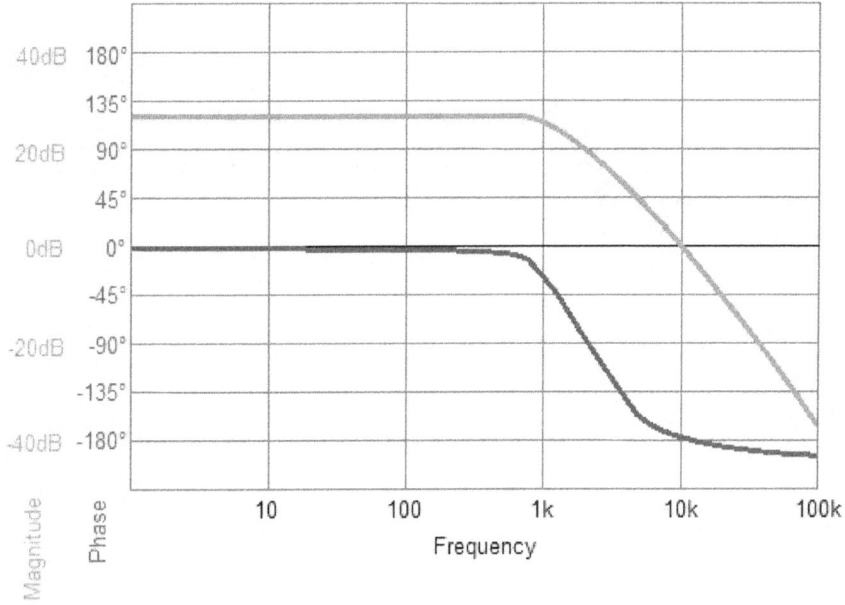

HW_A29:
Bode plot.

Bode plots are very popular for displaying the frequency characteristics of filters and controllers (such as a voltage regulator, which is really a control system attempting to control the output voltage of the regulator).

Bode plots are an invaluable source of information, displaying both magnitude and phase simultaneously over a logarithmic frequency axis.

There are many good sources available for learning about Bode plots and their use in discussing voltage regulators in particular, but one that I have found useful is: ON Semiconductor *DC-DC Converters Feedback and Control*, freely available online.

HW_Q30:

ssuming the previous Bode plot represents a control system, do you see anything potentially wrong?

HW_A30:

This control system is very likely to be unstable and fall into oscillation. The phase margin is close to 0°!

There are two major pieces of information that you should be able to quickly determine by looking at the Bode plot of a control system.

One is the frequency response of the system. In this case, we can see that the gain quickly falls off after about 1kHz, meaning the system will not be responsive enough to provide control above this frequency.

The second is the relative stability of the system based on its phase margin. The phase margin is the phase difference from -180° when the gain is 0dB (or 1). A low phase margin is a bad thing. In the diagram above, we can see that when the gain is at 0dB, the phase is almost exactly -180°, meaning the phase margin is close to 0°. This is perfect for an oscillator, but not for a control system. The output will respond to a disturbance at this frequency with an amplitude and delay to cause exactly the same amount of error in the opposite polarity, which will than cause a response to this error with an amplitude and delay to cause exactly the same amount of error in the opposite polarity, which will then... (you get the picture). If a control system is capable of oscillating, it is almost guaranteed to do so. It is amazing how quickly poorly designed control systems are able to turn themselves into oscillators, and so understanding this type of analysis is an important skill for control system engineers to possess.

A similar though less often used measure of stability is the gain margin. This is the gain difference from 1dB when the phase is -180°.

Lab Questions

L_Q1:

Tell (or show) me how to measure the voltage of this AA battery using this multimeter.

L_A1:

1) Switch the multimeter to DC Voltage mode. If multiple voltage ranges are available (most newer multimeters have auto-range selection and so this is rarely necessary), select the lowest range which will allow for reading the maximum expected voltage (lower ranges offer greater resolution). A AA battery has a typical voltage of 1.5V.

2) Ensure the multimeter probes are configured correctly. Most multimeters feature three or four probe inputs (ports). A common (typically colored black) port is used for all measurements. A voltage / resistance (typically red) port allows voltage and resistance measurements. One or more current (also typically red) ports allow current measurements. The purpose for multiple current ports is to allow greater resolution for lower current ranges. Some common multimeters, for example, have a current port that allows for measurements up to 400mA and a second current port that allows for measurements up to 10A.

In this case, we're measuring voltage, so make sure the positive probe is in the voltage port.

3) Place the negative probe on the negative side of the battery and the positive probe on the positive side of the battery and read the corresponding voltage.

This should be fairly straight forward. The interviewer is looking to see that you have some hands on experience with what ends up being one of a typical electrical engineer's most used tools. You should be able to easily pick up a multimeter and measure voltage, resistance, and current of circuits you will be working on.

L_Q2:

Tell (or show) me how to measure the period of the 10V peak-to-peak sine wave being generated by this function generator using this oscilloscope.

L_A2:

1) Connect the oscilloscope probe to the function generator, careful to clip the ground and signal leads to the appropriate terminals of the function generator.

2) Adjust the oscilloscope's Vertical control to allow for viewing the entire signal (+/-5V around the 0V horizontal line).

3) With the signal visible, enable the Trigger function and ensure its threshold position is within the signal range, preferably towards the middle of the signal.

4) Adjust the Horizontal control to bring the sine wave into clear view.

5) Use the Cursor function to measure the period of the sine wave.

Some new digital oscilloscopes will automatically measure and display the period and frequency of a repeating signal when the Trigger function is activated, and so the last step may not be necessary.

The oscilloscope is an invaluable tool that just about every electrical engineer will come to depend on time and time again, and so it is important that you have a good grasp of how to use this tool.

Some more experienced engineers may take secret delight in complaining about how younger engineers do not know how to use an analog scope (which are still abundant in many labs) so if possible to get your hands on, try to play around with an analog oscilloscope prior to interviewing. In reality, analog oscilloscopes are generally easier and more intuitive to use than their digital counterparts, even if they are not as functional.

At a minimum, you should be able to quickly identify and use the Vertical and Horizontal controls on any scope, as well as the Trigger function.

Tektronix, the electronic test equipment manufacturer, has many good oscilloscope related app notes and tutorials available on their website, such as *Oscilloscope Fundamentals*. The basic functionality of oscilloscopes from manufacturer to manufacture is relatively similar, so which specific brand of oscilloscope a company uses is not terribly important.

L_Q3:

What is an isolation transformer? Give an example of when it is appropriate to use one.

L_A3:

An isolation transformer is used to isolate mains power from Earth ground, typically for safety purposes in the lab. It is usually a 1:1 transformer, so 120 VAC in results in 120 VAC out. A good example for use is when measuring voltages on a mains powered DUT (Device Under Test) with an oscilloscope.

Typically, the grounds for all the channels on an oscilloscope are tied together and tied to Earth ground through the oscilloscope's power cord. A ground loop between the mains powered oscilloscope and mains powered DUT, therefore, could result in high currents that cause significant damage to one or both devices. Working around Earth referenced mains voltages also poses an electrocution risk, as your body can create a ground return path. The isolation transformer increases safety by reducing the likelihood of unintentional ground paths.

Note that some people mistakenly place the isolation transformer between mains and the oscilloscope. This is the wrong approach because it does nothing to limit electrocution risk when working with the DUT, and its effectiveness at reducing ground loop currents can accidently be circumvented when another grounded device is connected to the oscilloscope, such as a computer through a USB cable. Isolating the oscilloscope also increases the risk of electrocution by allowing high voltage potential to accidentally be applied to the probes' BNC connector shroud and other metal surfaces of the oscilloscope.

Software Questions

Not all electrical engineering interviews will include technical questions related to embedded software (or firmware) development. As the electronics industry continues to march towards digital microcontroller based designs, however, the need to be familiar with firmware development is becoming a greater necessity for electrical engineers.

Being familiar with firmware development will give you a significant advantage to others who are not and offers you a significantly greater number of employment opportunities. To date, the majority of embedded software projects are written in the C programming language, and so the de-facto embedded software test is geared towards knowledge of C.

C is also a great base language to know because many other languages are built around similar syntax and because it provides a solid background to how low level memory management works before learning higher level languages that take care of some memory management issues on their own (such as with garbage collectors).

That being said, being knowledgeable in additional programming languages, including higher level languages, is a benefit that will certainly add value to your prospects as a potential employee, but it rare that you will ever be expected to answer software language questions on any language other than C for entry level electrical engineering positions.

Embedded Software Engineer as a Career Choice

If you enjoy programming microcontrollers, you may want to consider embedded programming job opportunities in addition to or instead of electrical engineering positions. In many small companies and even in some larger companies, an electrical engineer will perform both of these roles at one point or another, but it is common for larger companies in particular to have specialized embedded software (or "firmware") engineers who focus primarily on writing software.

Lacking a software degree should not be a deterrent to embedded software positions because as is, the majority of the embedded software engineers have electrical or mechanical engineering backgrounds. And as university software engineering programs seem to be focusing ever more on higher level programming languages such as Java, education in lower level programming in C seems to have fallen by the wayside. Many new software engineering graduates, therefore, struggle to answer even some basic

---

embedded software interview questions. In addition, it is a common belief that it is easier to train a software savvy electrical engineer to be an embedded software guru than it is to train a high level software guru to be an embedded software guru. You already have a solid electronics background and lab skills (such as knowing how to use an oscilloscope) and polishing your software skills is something that makes for easier self-study material than electrical engineering principles. Having controls knowledge is also a big plus for a potential embedded candidate as embedded software engineers are commonly programming loops to control motors, temperature controllers, etc., and is why mechanical engineering is another common pathway to embedded software engineering.

Market demands are always changing, but currently, it is difficult to find high quality embedded software engineers and as the world of software does not appear to be going away anytime soon, embedded software engineering positions should be taken into consideration.

SW_Q1:
What is the number 10 in hexadecimal notation?

SW_A1:

A

Hexadecimal representation is commonly used in embedded programming. There is no need to be able to read large hexadecimal numbers and convert them to decimal in your head (that is what calculators are for) but you should have a general concept of what hexadecimal representation is.

```
    Decimal: 0 1 2 3 4 5 6 7 8 9 10 11 12 13 14 15 16
Hexadecimal: 0 1 2 3 4 5 6 7 8 9 A  B  C  D  E  F  10
```

One nice thing about hexadecimal format is that it allows the representation of a byte with two characters. A byte (8-bits) is capable of representing 0 through $(2^8 -1) = 255$ in decimal, or 0x00 through 0xFF in hexadecimal.

You should also know that numbers represented in hexadecimal format are commonly preceded by 0x in many programming languages and engineering documentation.

SW_Q2:
What is the number 5 in binary?

SW_A2:
101

Although it is not very common to actually use binary representation in embedded firmware (reading anything other than very small numbers in binary will quickly give you a migraine and the C programming standard technically does not offer binary representation support anyways), having basic knowledge of binary representation is important for many aspects of embedded programming, such as reading or writing to microcontroller control registers.

Likewise, it is a good idea to be familiar with a short power of 2s table.

$2^0 = 1$
$2^1 = 2$
$2^2 = 4$
$2^3 = 8$
$2^4 = 16$
$2^5 = 32$
$2^6 = 64$
$2^7 = 128$
$2^8 = 256$
$2^{10} = 1024$
$2^{12} = 4096$
$2^{16} = 65$ thousand and something... (memorizing the full value will clearly identify you as a geek)
$2^{24} = 16$ million and something...
$2^{32} = $ ridiculously huge...

Knowing the low 2s power values (0-8) will allow you to convert small numbers (such as anything that can be represented by a byte) to decimal, and vice versa. To solve the above question, for example, we can see that we can create 5 by summing 4 and 1, to give us a binary value of 101 ($1x2^2 + 0x2^1 + 1x2^0$) = (4 + 0 + 1) = 5.

Knowing the 2s power values for 8, 10, 12 and 16 will allow you estimate the numerical resolution of common ADCs and DACs. Note that the effective resolution may be lower than the numerical resolution and will depend on the specifications of the ADC/DAC and its implementation in your specific application.

Knowing the 2s power values for 8, 16, and 32 allow you to estimate whether or not an 8 bit, 16 bit or 32 bit word will be sufficient for holding a given piece of data. This is extremely important for embedded programming. If, for example, you need to store an integer value with a range between 0 and 100, one byte will suffice. If you need to store an integer value with a range between 0 and 1,000, on the other hand two bytes are necessary.

Luckily, memorizing all the previous values is not actually necessary for lower value powers. As is clearly evident, you just need to remember that the value doubles for each incremental increase in the power.

1000010110101...

For a quick mental break, check out the robot binary solo in the Flight of the Conchords song, *The Humans are Dead*, on YouTube.

SW_Q3:
What are the keywords `volatile` and `static` used for in the C programming language?

SW_A3:

There are multiple uses for each:

`volatile`

1) `volatile` can be used to tell the compiler that a variable can be modified by hardware (external to the software), e.g. a microcontroller memory mapped register holding the result of an analog to digital conversion or by other software called inside an interrupt routine.

2) `volatile` can be used to prevent the compiler from optimizing away operations on a variable, such as the continuous incrementing of an integer to implement a software delay.

In short, `volatile` is used whenever we want to force the compiler to read the value of a variable every time it is referenced, and to prevent it from storing a previous value in memory or optimizing away operations on that variable. It is important to note, however, that using software delays is generally frowned upon. It is preferable to use a hardware timer peripheral if possible.

`static`

1) `static` can be used to declare that a variable inside a function definition should retain its value between function calls.

2) `static` can be used to declare that a function or variable (outside of any function) is local to the *.c file it is declared in, which prevents functions in other files from calling this function or accessing this variable.

For 1), it should be understood that variables inside a function are created (in memory) when a function is called and destroyed when the function is exited. Typical function variables, therefore, are not able to retain their value between function calls. They will be reinitialized each call. The `static` keyword allows you to specify that the variable should remain in memory across function calls. The advantage to using a `static` variable inside a function, as opposed to a global variable, is that the `static` variable prevents access by anything outside the function, and so provides some protection. In general, it is considered bad programming practice to use global variables and good practice to limit the scope of variables as much as possible.

For 2) the use of `static` helps promote good software design principles, such as data hiding and abstraction, by limiting the scope of variables and functions to a single module (*.c file containing functionally related data and functions). If writing a module to interface to an external EEPROM, for example, you may have `read_byte(int address)` and `write_byte(int address)` type functions included in the interface, but keep lower level EEPROM communication functions `static`. The idea is to implement "black box" type functionality, where only the high level interface is exposed, and the low level details are hidden and kept out of reach of the end user.

Explanation requests for the `volatile` and `static` keyword use are very common in embedded software interviews.

An absolute must study reference that addresses these keywords; along with other embedded C questions is: *A 'C' Test: The 0x10 Best Questions for Would-be Embedded Programmers* by Nigel Jones (freely available on the Internet).

This article was originally published in *Embedded Systems Programming*, a magazine which no longer exists in print, although carries on as a shadow of its former self at www.embedded.com. Despite its enormous popularity, the questions from this article continue to be common in interviews, either because so many younger engineers have not been exposed to it or because embedded programming engineers coming up with interview questions tend to think alike.

Pay special attention to the question related to Bit Manipulation, as questions on how to set and clear bits in C using the logic & (AND) and | (OR) operators are very common in embedded software interviews.

SW_Q4:

Can you explain the difference between the heap and stack memory regions?

SW_A4:

The stack is an area of memory used to hold local variables and related function data when a function is called. This data is automatically released after the function has been processed. This acts as a LIFO (last in, first out) buffer. If function A calls function B which in turn calls function C, for example, all three functions will be simultaneously on the stack. Once C returns a value to B it is released from the stack, and then when B returns a value to A it is released from the stack. "Stack overflow" refers to a fault that occurs when there is not enough stack memory left to hold additional stack data. This usually leads to catastrophic failure.

The heap is an area of memory that holds dynamically allocated data. To allocate data in the heap, software must specifically request heap space, such as with the `malloc()` function. Software must also manually free heap data when it is no longer needed, such as with the `free()` function. As heap data is dynamically allocated and freed, there is no standard queue arrangement to this data.

The use of dynamic memory allocation is generally frowned upon in embedded programming (something that you should mention if discussing this subject). It is much more difficult to prove (either with static analysis or testing) that a product will be stable over a long time period with a given sized heap than it is with a given size stack. As many embedded products are used in applications that call for high reliability and possibly extremely long run times with very limited resources, this can be a problem. Even a large heap can suffer from fragmentation. This can result in an allocation failure when there is not enough contingent area of the heap available for allocation, even when a significant percentage of total heap space may be free. This may occur when many small blocks of data are blocking a larger block of data from being allocated. Use of the heap also puts a larger burden on the programmers, who are responsible for ensuring that data is allocated and freed properly. If data is allocated periodically but never freed (a memory leak), it is just a matter of time before the heap becomes full and an allocation fails.

To avoid these issues, many embedded systems avoid the use of dynamic memory allocation altogether, or use it within strict confines, such as only one time at system initialization. This is especially true for safety or reliability critical applications.

SW_Q5:

Can you see a potential problem with the following function?

```
/*****************************************************
 * This function converts an integer temperature
 * in degrees C to degrees F.
 *****************************************************/
int8_t convert_temp_c_to_f(int8_t  temp_deg_c)
{
   int8_t  temp_deg_f;

   // F = C*(9/5) + 32
   temp_deg_f = temp_deg_c*9/5 + 32;

   return(temp_deg_f);
}
```

SW_A5:

Floating point operations can be expensive on microcontrollers without a hardware floating point unit (FPU), which most low cost microcontrollers do not have. Microcontrollers without floating point units can still perform floating point operations with a floating point library, but it may take a significant number of processor cycles to perform common floating point operations and the floating point math libraries can consume a significant amount of precious program memory (Flash). For these reasons, floating point operations are commonly avoided when unnecessary in embedded applications, specifically those using older or low cost, low performance microcontrollers.

Along these lines, the function in this question performs an integer (non-floating point) conversion in temperature from degrees Celsius to Fahrenheit (note the int8_t type definition from `<stdint.h>` used to specify a signed byte type for both the function parameter and return value).

It is assumed that for this particular application, precision to less than a whole degree is not required (e.g. we only need to know that the temperature is approximately 22°C, and not that it is 22.6°C. Although this may be assumed for this question, it is worthwhile discussing while answering the question, along with how greater precision can still be achieved using fixed point arithmetic (see below).

A potential problem with this function as written is that too large a temperature will cause an overflow condition that will result in a grossly inaccurate conversion. The range of a signed byte (using two's complement) is from -128 to 127 and an unsigned byte is from 0 to 255.

As the above conversion function multiplies the passed temperature by 9 before dividing by 5 (multiplication and division have equal precedence and therefore are evaluated left to right), any passed temperature greater than 29°C will cause an overflow during the multiplication operation (9x29 = 261, which is greater than 255).

To increase the temperature range while keeping the passed and returned variables unsigned bytes, the passed temperature can be cast as a 16 bit word during the multiplication step.

```
temp_deg_f = (int8_t)((int16_t)temp_deg_c*9/5) + 32;
```

In this case, the multiplication and division occur with 16 bit words, so that no overflow is possible (the highest possible temperature to pass to the function is 127, and 127x9 = 1,143, which is well below the +/-32 thousand and something... limit of a signed 16 bit word.

Now, the overflow limit worry is dictated by our signed byte return value (+127 / -128). This limits the valid Celsius temperature input range to:

(127-32)5/9 = 52
(-128-32)5/9 = -88

-88 through 52 degrees Celsius. At the very least, it would make sense to limit the calculated Fahrenheit value to within this range before casting it back to a uint8_t type to prevent a completely bogus value from being returned.

Any function that performs arithmetic that can cause errors, such as overflows or divide by zero errors, should limit check and prevent these conditions from occurring. This is commonly referred to as "defensive programming." Likewise, a note in the function's comment section should specify any limitations the function has, including if the range of a parameter is not the same as data type it uses.

Fixed Point Arithmetic

While in many applications it may be desirable to avoid using floating point representation, something greater than integer resolution may be required. Fixed point arithmetic is commonly used to allow some fractional resolution without the demanding resource requirements of floating point operations.

This can be accomplished simply by multiplying by a certain decided upon value to achieve the desired level of resolution. If we wanted to represent temperature with 0.1° of resolution using an integer type, for example, we could simply multiply the temperature by 10. 22.3°C, therefore, would be represented by 223. When dealing with fixed point values, it is necessary for all functions working with those values to be aware of the multiplication factor, and various variable naming conventions are used to help make it more obvious to the programmer when fixed point math is used.

It is common for fixed point arithmetic to use multiplication factors that are powers of two, because multiplying and dividing by powers of two can be done very efficiently in most microcontrollers by simply shifting left or right (dividing by four, for example, can be accomplished with two right

---

shifts).

Standard Integer Types

When reading embedded C software, it is common to see integer types defined using exact-width integer types as defined in the `<stdint.h>` header file (starting with C99). For example, `uint8_t` can be used in place of `unsigned char`. This makes the code more portable as the actual width of standard c types can be system dependent (e.g. an `int` in a 16 bit system may be defined differently from an int in a 32 bit system).

SW_Q6:

Can you see a potential problem in the following code?

Assume that the microcontroller ADC is configured to automatically sample and update the `adc_value` at 1,000Hz with `adc_interrtup()`, and that the `transmit_adc_value_to_host()` function is called by the main application to transmit the most recently sampled ADC value when requested by the host.

```
#define ADC_OFFSET 4

static volatile U16_T adc_value;

pragma __interrupt__
adc_interrupt()
{
    clear_interrupt_flag();
    adc_value = ADC_SAMPLE_REGISTER;
}

void transmit_adc_value_to_host(void)
{
    U8_T transmit_buf[2];

    transmit_buf[0] = (U8_T)(adc_value & 0x00FF);
    transmit_buf[1] = (U8_T)((adc_value >> 8) &
0x00FF);

    transmit_to_host(transmit_buf, 2);
}
```

SW_A6:
The `transmit_adc_value_to_host()` function is not reentrant, meaning an issue can arise if the function is interrupted.

For example, imagine the following scenario:
assume that `adc_value` = `0x01FF` from the last ADC sample.

The main application calls `transmit_adc_value_to_host()` which starts to load `0x01FF` into the transmit buffer.

First, `transmit_buf[0]` is set to `0xFF`.
```
transmit_buf[0] = (U8_T)(adc_value & 0x00FF);
                   (0x01FF    & 0x00FF) = 0x00FF
```

Next, the `adc_interrupt()` occurs, and the `adc_value` is updated with a new sample value of `0x0200`, corresponding with a small increase in the analog input voltage since the previous sample. The interrupt completes and the `transmit_adc_value_to_host()` function continues where it left off, now loading the final `adc_value` byte into the transmit buffer. The problem is that the transmit buffer is now loaded with high and low bytes from two different samples.

```
transmit_buf[1] = (U8_T)
                   ((adc_value >> 8) & 0x00FF);
                   ((0x0200    >> 8) & 0x00FF) = 0x0002
```

The end result is the transmission of `0x02FF`, which is drastically different than either of the `0x01FF` or `0x0200` sample values that led to its creation.

One way to address this issue is to surround the buffer load statements with calls to disable and enable the interrupt. This prevents `adc_value` from being changed while it is loaded into the buffer. Disabling interrupts can negatively impact performance and even lead to errors, which is why this sort of technique should not be used around large sections of code. If a communication peripheral interrupt, for example, was used to empty a receive buffer when full, disabling interrupts for too long could lead to incoming data over-writing old data that had not yet been processed.

```
Disable_Interrupts();
transmit_buf[0] = (U8_T)(adc_value & 0x00FF);
transmit_buf[1] = (U8_T)((adc_value >> 8) & 0x00FF);
Enable_Interrupts();
```

Another way to address this is to store the value of adc_value into a temporary value.

```
void transmit_adc_value_to_host(void)
{
    U16_T temp_adc_value;
    U8_T transmit_buf[2];

    temp_adc_value = adc_value;

    transmit_buf[0] = (U8_T)(temp_adc_value & 0x00FF);
    transmit_buf[1] = (U8_T)(( temp_adc_value >> 8) &
                          0x00FF);

    transmit_to_host(transmit_buf, 2);
}
```

This works assuming that the assignment of adc_value to temp_adc_value is atomic, meaning it executes in one clock cycle and so cannot be interrupted. When working on interrupt software such as this, it pays to have a deep understanding of the microcontroller hardware and compiler output to know what operations are atomic and what are not. For code that must be reentrant, assumptions should not be made.

Programming interrupts is something that is unique and common to embedded software work, so being familiar with their operation and potential pitfalls will give you an advantage over many other entry level embedded software engineers.

A good article to read about the potential pitfalls to embedded software is *Five top causes of nasty embedded software bugs* by Michael Barr. This article is freely available on the www.embedded.com website.

SW_Q7:
Write a function to debounce a momentary push button input and return 1 to indicate the pressed edge of the button, 0 otherwise. Assume this function is called at a regular interval, say 100 times a second.

SW_A7:

First, you must understand what button "debouncing" is and why it is required.

When a mechanical button is pressed, its contact will "bounce" and cause the resulting voltage to fluctuate high / low for a period of time before the voltage settles on its final state. If a processor were to read the state of the button at a fast rate, a single button press may appear as multiple button presses. This can result in undesired behavior. A poor or missing button debouncing algorithm, for example, might result in a single press of the "channel up" button resulting in the channel incrementing two or more times.

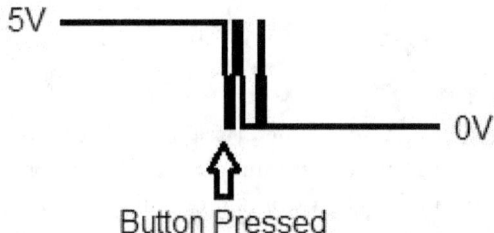

Button Pressed

Interfacing with buttons and switches is very common in embedded applications, and this sort of question goes along with the idea of asking about real world engineering problems that may not necessarily be dealt with in school.

It helps to visualize what a momentary button press might look like at the digital input of a microcontroller. Assuming a pressed button reads low or 0 and a non-pressed button reads high or 1, a button press may look like the following.

```
1111111110101100110100000000000000000011011111111111
         ^          ^                  ^ ^
         a          b                  c d
```

Here, the button is pressed at sample c and bounces until sample b. It is then released at sample c and bounces again until sample d. As is clearly visible, simply looking for the transition from 1 to 0 would lead to many false button press indications. In the above example, a single button press would result in 6 pressed edge indications, 5 of which would be false.

What we want to do is look for a transition from 1 to 0 followed by multiple consecutive 0s. You can see an obvious trade off here between detecting false button presses and missing button presses entirely. If the required number of consecutive 0s is too large, the button presser may have to learn to hold down the button for an excessive amount of time to ensure his presses are read, and he will likely not have a pleasant user experience.

All this should be discussed with your interviewer to show some understanding of debouncing and to demonstrate that you are not simply picking numbers out of thin air. You may decide that 5 consecutive samples, or 50ms of a clean pressed signal, should be a good balance between button response and bounce susceptibility, but stress that in "real life" you would carefully analyze the performance of sample buttons under use and thoroughly test your choice.

From here, you need to write a function that will return true if it sees the button state change from not pressed to pressed and followed by 4 more consecutive pressed states. Just as important as writing a functionally correct function is demonstrating that you have good software development habits and will not be torturing your future coworkers with hard to read and maintain code. You should do this by using #define or const instead of magic numbers, and comments where necessary.

One thing you don't need to be overly worried about is small syntax errors. If you miss a semicolon somewhere but do everything else well, it is doubtful the interviewer will hold it against you. Compilers and the Internet are always available to quickly solve small syntax errors in real world development. Many times, you will simply be asked to write pseudocode without regard to the detailed syntax requirements of any given language, especially if you are up at a whiteboard.

This function can be implemented many different ways, but below is one particularly popular example implementation.

```
//The below defines assume a pressed button I/O state
// of 0, and require 5 consecutive button pressed
// states for pressed edge detection.
#define BUTTON_STATE_MASK    (0x3F) //111111
#define BUTTON_PRESSED       (0x20) //100000

//Returns 1 if a pressed button edge is detected,
// else 0.  Takes the raw button I/O value at 100Hz
uint8_t button_pressed_edge(uint8_t raw_button_io)
{
```

```
    static uint8_t button_state = 0xFF; //not pressed

    button_state = button_state << 1;
    button_state = button_state | raw_button_io;
    button_state = button_state & BUTTON_STATE_MASK;

    if (BUTTON_PRESSED == button_state)
    {
        return(1);
    }
    else
    {
        return(0);
    }
}
```

Two #defines are used to set up the function. These help eliminate the use of magic numbers to make the code more readable and make it easier to adapt the function in the future, say to change the number of consecutive raw button pressed states required to trigger an edge detection condition. It may be possible to use some macros to make the function even easier to modify in the future, such as by using one #define to declare the (high/low) state of a button press and a second #define that sets the number of consecutive raw sampled pressed states required for a pressed edge detection. In the interest of time, however, you should not get too carried away with these sorts of things. They may make for good interview conversation, but demonstrating understanding of the problem and the ability to engineer a working solution is the top priority.

The function itself is fairly fast and memory efficient (generally a good thing for embedded systems). It works by shifting the current state of the button input into a static button_state variable. The fact that this variable is static means that it will retain its value between function calls. In effect, it acts as a running history of the last 8 button samples. Another nice thing about this function is that it demonstrates some knowledge about the bitwise operators | (OR) and & (AND) as well as the left shift operator <<.

The classic guide to debouncing inputs is *A Guide to Debouncing* by embedded guru Jack Gannsle, available for free on his website (www.ganssle.com).

SW_Q8:

Write a function that will swap the values between 2 integer variables. Then write a statement calling that function to swap two integer variables, x and y.

SW_A8:

```
void swap(int *a, int *b)
{
    int temp;

    temp = *a;
    *a = *b;
    *b = temp;
}


swap(&x, &y);
```

This is a classic C question used to evaluate your knowledge of pointer operations. As college computer science courses have gradually moved to higher level languages such as Java, this question seems to have become more difficult for new computer science graduates who may be applying for an embedded software position.

It is important to understand why the below function does not work.

```
void bad_swap(int a, int b)
{
    int temp;

    temp = a;
    a = b;
    b = temp;
}
```

When this function is called, such as:

```
bad_swap(x, y);
```

The value of x is copied into a and the value of y is copied into b. The values are then swapped between a and b, but x and y remain the same. When the function completes, a and b are destroyed (their memory location is released from the stack for other functions to use) and the values of x and y remain the same.

```
bad_swap(&x, &y);
```

The above function call doesn't help. In this case, instead of copying the

130

values of x and y to a and b, the memory addresses of x and y are copied into a and b. The address values are swapped between a and b with no change to the values in x and y.

The correct solution uses pointers to allow direct manipulation of the values stored at the memory locations passed to the swap function.

```
swap(&x, &y);
```
copies the address of x and y into a and b

```
temp = *a;
```
copies the value located at the address a into temp

```
*a = *b;
```
copies the value located at the address b over the value located at address a

```
*b = temp;
```
copies the value of the temp variable over the value located at address b

Pointer operations can be a bit tricky if you're not familiar with them, and so they make a good subject to review prior to any interview where you believe embedded programming questions are likely.

SW_Q9:

What can you tell me about this microcontroller block diagram?

See block diagram page of any common microcontroller datasheet. Good examples can be found for the microcontroller products from companies such as Microchip, Texas Instruments, NXP, and STMicroelectronics.

SW_A9:
Almost every microcontroller datasheet will include a block diagram, usually towards the beginning of the document. This will include a high level view of the microcontroller, including its integrated peripherals. Asking a candidate to share his knowledge of such a block diagram is a good way to get an understanding of his familiarity with microcontrollers.

Common Microcontroller Components

Processor - In the embedded world, 8, 16 and 32 bit processors are commonplace, though ever improving semiconductor process technology continues to shrink the price and low power advantages of 8 and 16 bit processors over 32 bit processors, and it seems that new designs are more likely to use 32 bit microcontrollers than in the past. One thing that the lower bit microcontrollers have going for them is momentum, as there is a significant number of legacy designs still using 8 bit microcontrollers and the software re-work that is required to port code to a newer microcontroller can be daunting.

The majority of embedded microcontrollers do not have hardware floating point units, although their use is seeing some rise with the ever forward march in semiconductor technology. The logic and arithmetic capability of microcontrollers varies wildly depending on the cost and performance of any particular microcontroller. In the past (when embedded software was written primarily in assembly language) engineers spent a great deal of time learning about different processors and their instruction sets. It was common for debates about various processors' strengths and weaknesses to occur during the component selection process. The Microsoft versus Mac versus Linux debates were replaced by the Intel versus Motorola versus Zilog debates. Today, many engineers still have their favorite embedded processors, but as most embedded code is now written in C or C++, a detailed understanding of a processor's instruction set is usually not necessary.

Data RAM - Data RAM (Random Access Memory) is the working memory of the processor.

Program Flash - Program Flash is non-volatile memory that the program is stored in. Most modern microcontrollers are capable of writing to Flash through software, although some require a higher voltage rail in order to accomplish this. Flash has a limited number of write cycles, however, and so it is not typically used in a repetitive way that could cause "wear-out" to occur within a products expected lifespan.

EEPROM - Electrically Erasable Programmable Read-Only Memory (EEPROM) is a rare component of modern microcontrollers. The same process improvements that have led to ever more advanced, faster, and cheaper microcontrollers have, unfortunately, left the internal EEPROM peripheral a relic of the past. But while most modern microcontrollers require an external, dedicated EEPROM IC, you will still occasionally see an internal EERPOM in older microcontroller datasheets. Some newer microcontrollers will have "emulated" EEPROM that uses a section of Flash designed for greater write cycles.

EEPROM is NVM (non-volatile memory) used to store information between power cycles (which would otherwise be lost in RAM) and typically is rated for hundreds of thousands of write cycles.

Note that some newer microcontrollers are starting to include FRAM (ferroelectric random access memory) NVM.

Memory Mapped Registers - Interfacing with memory mapped registers is a common task for embedded software engineers. Memory mapped registers provide an elegant way for software to interface with hardware peripherals. Setting or clearing bits in registers at specific addresses allow the software to set an I/O pin high or low, turn a PWM output on or off, etc.

As bit-wise operations on memory mapped registers are so commonplace, you should be familiar with common operations to set and clear individual bits, say for example, to toggle I/O pins.

To set a bit in a register without changing any of the other bits in that register, the register should first be read into a temporary variable. This temporary variable should then be OR'd with the bit to be set, and then written back to the register location.

To clear a bit in a register without changing any of the other bits in that register, the register should first be read into a temporary variable. This temporary variable should then be AND'd with the bitwise complement of the bit to be cleared (e.g. to clear bit 3 of an 8 bit register, AND with 11110111, which is the equivalent to ~00001000), and then written back to the register location.

It is important to be mindful of read-modify-write operations on registers that could change value during the process. It may be necessary to disable interrupts or take other precautions to prevent data corruption (see previous comments on reentrant code in SW_A6).

Some microcontrollers offer special registers to set or clear bits more easily. This functionality is common with output pin control registers, for example, that allow setting an output pin high or low directly with special registers that do not require performing a read-modify-write operation.

Interrupt Controller - Working with interrupts on a regular basis is one of the distinguishing features of embedded programming. Interrupts allow fast, real time response to events.

The interrupt controller may allow priority settings of individual interrupts, so that a higher priority interrupt can interrupt a lower priority interrupt.

GPIO - General Purpose Inputs / Outputs are the simplest of microcontroller peripherals, allowing for basic software interfacing to external circuitry. Most microcontroller GPIO pins can be configured as either inputs or outputs. Some pins may also have configurable pull-up / pull-down resistors and may be configured as either push-pull or open-collector when used as an output.

A push-pull output will drive a logic low to ground or logic high to the I/O voltage supply rail, while an open collector output will only drive low, and allow the pin to float for the logic high condition, in which case a pull-up resistor is used to pull the voltage up to the logic high voltage level.

Open collector outputs can be useful for interfacing two components that use different logic high voltage levels, though care should be taken to ensure the maximum voltage for both devices is not exceeded. Open collector outputs are also commonly used for certain communication schemes in which multiple devices share control over a single line. A single pull-up resistor brings the line high, and any of the devices sharing the line can pull it low. If a "collision" is to occur when multiple devices simultaneously pull the line low, no damage will occur as the maximum current is limited by the pull-up resistor. If multiple devices share a communication line with push-pull outputs, on the other hand, damage may occur when one device attempts to drive the line high while another attempts to drive it low (creating a short circuit condition).

By default, most microcontrollers configure their pins to inputs, and must be purposefully configured to outputs before they can be used as such. This is to prevent damage to external circuitry from occurring during power up and before the I/O is configured. As previously described, two output pins fighting with each other can cause damage, while damage is much less likely to occur from two connected inputs.

Defaulting I/O to inputs, however, is not guaranteed to prevent undesirable behavior at power-up and before pins are initialized. A device may enter an undesirable state if one of its inputs is not driven low or high. To account for this, it is common to use weak pull-up or pull-down resistors on inputs to ensure that a default "safe" state exists before the microcontroller completes its initialization process.

In addition to a functionally safe state, pull-up / pull-down resistors prevent input voltages from floating between logic thresholds when no signal is present. A voltage that floats between low and high voltage logic thresholds on a digital input can potentially engage both the low and high side input sensing transistors, creating a partial short circuit which can overheat and damage the input. Newer microcontrollers may include Schmitt trigger protection on inputs to prevent this from happening, but older digital devices can be susceptible to this kind of damage.

Care should be taken to note what internal pull-up / pull-down resistance inputs default to, if any, to prevent undesirable interaction with external resistors. If, for example, a 100kOhm external pull-up resistor is used to set a digital device input to a safe state, while a default internal pull-down resistance of 100kOhm is incorporated in the microcontroller I/O connected to the same line, the voltage at the input pin will be half the logic voltage supply, resulting in a potentially undefined state that could lead to wasted power or even input damage. If the default pull-down resistor configuration is known, a stronger pull-up resistor can be used, say 10kOhm, to ensure the input voltage meets the required logic threshold conditions before the microcontroller software has time to initialize the pin.

Timer - Timer peripherals perform basic timing functions which can be important for real time embedded systems. They can be configured to implement accurate delays or to pace the period of an ADC sample rate, control loop, or other time critical events. Timer peripherals can also be used to accurately measure the time of a given process or event. Most timer peripherals include some interrupt capability to trigger an interrupt after a specified amount of time has passed.

PWM - Pulse Width Modulator (PWM) peripherals are used to generate PWM outputs, which are commonly used to control average current such as in switching power supplies, LED brightness control applications, and motor control applications. They are also used to control average voltage such as with a makeshift DAC by using a RC filter. PWM modules typically include their own timer module or use one of the microcontroller's existing timer peripherals and may also include some application specific features, such as for brushless motor control.

Input Capture - Input Capture functionality may be included in a Timer module or may exist as its own peripheral. This peripheral can be used to capture the Timer value when an input pin transitions. The Input Capture peripheral can be configured to catch High->Low transitions, Low->High transitions, or both, and can also typically be configured to time-stamp each transition or after a certain number of transitions (e.g. after 8 transitions). This module is useful for measuring the period of an external PWM signal or interfacing with an encoder or tachometer.

Output Compare - Output Compare functionality mirrors that of Input Capture. An output pin can be programmed to transition at a certain Timer value. Different raising and falling edge transition time values can be used to implement PWM functionality.

I²C - Inter-Integrated Circuit (I²C) is a serial communication protocol that allows multiple devices to communicate over a two wire bus. The two communication lines, SDA (Serial Data Line) and SCL (Serial Clock Line) are pulled up to a logic-high voltage through resistors and all I²C devices connect to these lines with bidirectional open-drain pins. This configuration allows for multiple masters to share a bus. I²C devices are identified on an I²C bus with a unique address (I²C addresses are distributed by NXP – formerly the semiconductor division of Philips and inventor of I²C).

I²C peripherals implement low level I²C logic to limit the amount of processing work required by the processor to transmit and receive information.

I²C is commonly used for connecting multiple low speed components together on a PCB while keeping the I/O requirement low. I²C bus speed is typically fairly low (100kHz-standard or 400kHz-fast), although some higher speed implementations exist. It is a popular interface to external EEPROM (Electrically Erasable Programmable Read-Only Memory), RTC (Real Time Clock), ADC (Analog to Digital Converter), DAC (Digital to Analog Converters), and digital sensors, such as temperature and MEMS (Microelectromechanical Systems) inertial sensors.

I²C is especially useful for buses that may have multiple masters, such as when multiple microcontrollers share a bus, and is also useful when a single bus has components that are powered from different voltage rails, although you must always ensure that the I/O for each device is compatible with the pull-up bus voltage.

SPI / SSI - Serial Peripheral Interface (SPI), or sometimes referred to as Synchronous Serial Interface, is a serial communication protocol that allows one master to communicate with multiple slaves. SPI communication allows the master and slave to transmit information simultaneously (full duplex).

Three communication lines are used for typical SPI configurations, SDI (Serial Data In), SDO (Serial Data Out) and SCK (Serial Clock). The SDI and SDO terminology can be confusing as the SDO of a master will be connected to the SDI of a slave and vice versa. For this reason, it is more common to label these lines in schematics as MOSI (Master Out Slave In) and MISO (Master In Slave Out), which allows a single net name to describe the function for both the master and slave.

A fourth line, SS (Slave Select) is used by the master to select the slave device. If multiple slaves share a common bus, a unique SS line is required for each of them. The total number of lines (and pins on the master device) therefore, is 3 + N (where N is the number of slave devices). If only a single slave device is used, a SS line may not be required and in this case, the SS pin on the slave device is simply pulled to ground (always selected). Some slave devices, however, require a falling edge of the SS line to proceed data transmission, in which case a SS line is required.

SPI peripherals implement low level SPI logic to limit the amount of processing work required by the processor to transmit and receive information. Different device manufacturers commonly use slightly different SPI protocols (or modes), and so most SPI peripherals include settings to specify which clock edge data is transmitted and latched on and whether the clock idle state is high or low.

SPI is commonly used for interfacing with many of the same types of components listed in the I²C section. Although SPI requires more lines than I²C (in particular for buses with multiple slaves), it generally offers faster speeds than I²C, with typical clock speeds in the 1-10MHz range and beyond.

UART / SCI - Universal Asynchronous Receiver/Transmitter (UART) peripherals, or sometimes called Serial Communications Interface (SCI) peripherals, are one of the most common microcontroller communication peripherals.

UART may be used at a board level to allow serial communication between two microcontrollers or other UART enabled device or it may be used in conjunction with a transceiver to implement off-board protocols such as RS-232 and RS-485. Typical transfer rates vary over a wide range, from 100s of bps (bits per second) to 10+ Mbps (Megabits per second).

UART peripherals implement low level UART logic to limit the amount of processing work required by the processor to transmit and receive information.

Some modern microcontrollers also include optional synchronous capability in their UART peripheral, and may therefore refer to their UART modules as USART.

CAN - Controller Area Network (CAN) is yet another serial communication bus. It was developed for the automotive industry to provide a robust, low cost communication interface between the various controllers in modern cars. It has since gained popularity outside of the automotive industry, and is now commonly found in industrial applications as well.

CAN uses a two wire differential bus and allows multiple masters to operate on the same bus. Typically sending small packets of information between "nodes" on the bus. CAN peripherals implement low level CAN logic to limit the amount of processing work required by the processor to transmit and receive information. Some microcontrollers' CAN peripheral will include CAN transceivers, allowing for direct connection to the CAN bus, though most require an external CAN transceiver.

Ethernet - Ethernet interfaces are becoming more popular in modern, high power microcontrollers. Various communication protocols can be used over the Ethernet data link layer, and require a fair amount of software. Many microcontroller companies offer free Ethernet stacks to use with their microcontrollers to get communication going without a large amount of custom software development work.

While there are a few microcontrollers that include a built in transceiver, or PHY (Physical Layer Transceiver), this is fairly rare. An external PHY chip is typically required, along with magnetics to isolate the two differential data pairs.

USB - USB is another common interface on modern microcontrollers and provides a convenient way to interface with most computers. Like Ethernet, there is significant software overhead, but many microcontroller companies offer free USB stacks to use with their microcontrollers. Unlike Ethernet, USB uses a simple electrical interface that in many cases requires little circuitry external to the microcontroller.

ADC - Analog-to-Digital converters provide a way for microcontrollers to interface with the analog word. The resolution and performance of ADCs built into most microcontrollers is typically lower than that of their external counterparts, but still offers sufficient resolution for many common analog sampling tasks.

DAC - Digital-to-Analog converters are a rare peripheral on microcontrollers, but are available in select parts. More commonly, for low resolution DAC functionality, a PWM or Output Capture peripheral can be used with an RC filter.

DMA - Direct Memory Access (DMA) controllers can be used to transfer blocks of data from one memory location to another. As large data transfers can be processor intensive, DMA controllers can be used to offload this task from the processor.

DMA controllers may be used to move data from one memory region to another or from a peripheral (such as UART, SPI, etc.) to memory or vice versa. One popular use of DMA controllers in more advanced microcontrollers that feature a LCD controller and external memory interface is to continuously refresh the LCD from its framebuffer (usually located in external memory), a task that would demand a significant amount of processor bandwidth without a separate DMA controller.

Watchdog Timer - A Watchdog Timer can be used to detect certain software and hardware faults. During normal operation, the software will periodically clear (or reset) the Watchdog Timer. If the software is to lock up or take an unpredicted logic path to a dead end, the Watchdog Timer will timeout and typically force a microcontroller reset.

For some applications however, a reset may not be the desired course of action when a timeout fault occurs (it may be more desirable to simply shut down until an operator can take further action, for example). In this case, most Watchdog peripherals will set a persistent flag that can be checked by software at initialization to determine if it just came out of a Watchdog reset.

Serial Programming / Debugging - Most microcontroller manufacturers include a serial programming / debugging port that allows for a simple, low pin count interface to the microcontroller.

JTAG - The Joint Action Test Group (JTAG) is the name for the Standard Test Access Port and Boundary-Scan Architecture standard. This was originally developed to facilitate inter-IC connectivity testing of dense circuit boards, but is commonly used in modern microcontrollers as a programming and debugging port as well.

Trace - More advanced microcontrollers may include a Trace port which allows for more sophisticated debugging and software profiling capabilities than with traditional debuggers.

Micro Lingo

Microcontroller / Microcroprocessor / DSP (Digital Signal Processor) / etc.

Microprocessors are bare bones processors. They typically require external RAM, Flash, etc.

Microcontrollers are essentially mini-computers. Along with a processor, they include RAM, Flash, and an assortment of various peripherals (as discussed above).

DSPs (Digital Signal Processors) fall into an ever blurring category. They are generally described as microprocessors or microcontrollers optimized for digital signal processing. Most modern designs using DSPs will have an architecture closer to the microcontroller, although some may require external program Flash. These devices typically include additional instructions and hardware capability for digital signal processing, such as a multiply accumulate (MAC) unit.

It is not uncommon to hear people call the same part any or all of the above terms, so while you should attempt to be as accurate as possible with your use of the terms during an interview, do not be surprised to hear them thrown around haphazardly by others.

As previously mentioned, some good source of DSP information:

The Scientist and Engineer's Guide to Digital Signal Processing, by Steven W. Smith, Ph.D. and freely available at www.dspguide.com.

Mixed-Signal and DSP Design Techniques, edited by Walt Kester and freely available on Analog Devices website (www.analog.com).

Other Types of Technical Questions

Troubleshooting Question

The interviewer or interviewers will typically provide you a schematic for a circuit or small system that has stopped working. They will then ask you what you would do (step by step), and what type of tools you would use to diagnose the problem.

One important thing to keep in mind for this type of question is that the problem will never be the first thing you check. The interviewer or interviewers will have a mental list of every possible problem, and will typically wait to see how many of them you can find, and not let you 'discover' the actual problem until there are no more potential problems to be discovered. This can be harder than it sounds, and it is not unheard of for the interviewer to lose track of what you've already checked. If this is the case, and the interviewer finally reveals that there was a faulty power supply, gently remind him that it was one of the first items you checked.

Keeping this in mind, if your roommate previously interviewed for the same position and eventually found that the troubleshooting question problem was with a faulty switch, you should not attempt to impress your interviewers by claiming you first step will be to replace the switch. Not only will you be wrong, but you will be demonstrating poor troubleshooting skills. Who replaces a part before testing it?

In addition to observing your troubleshooting skills, this type of question is designed to see if you can read and understand a basic schematic, and if you know about common electrical testing tools (such as a multimeter) and how to use them (e.g. "I place the red lead here and the black lead here and measure the DC voltage").

This knowledge should help you in three ways.

First, don't panic or get anxious if you don't find the problem right off the bat. Your goal is to demonstrate a thought out, methodological approach to pinpointing the problem, not to guess the right answer. Your troubleshooting process is what is under inspection. Try to partition the problem gradually into smaller sections of the circuit when possible. And as always, THINK OUT LOUD!

Second, don't try to pick the most obscure potential issue first with the false belief that it is most likely to be the source of failure or to impress your interviewers. This will backfire on you because it obviously will not end up being the problem and the interviewer will think it strange, possibly wondering if your troubleshooting process is based off random guessing. Don't out-think the question. Good troubleshooters will suspect the most obvious issues first. "Is it plugged in?" is not only easy to check, it is more likely to be the issue than some obscure semiconductor failure.

Third, don't get down on yourself after you finally find the problem and think 'I should have looked at that first!' There is no reason to allow these types of questions to start an avalanche of negative thoughts. With the knowledge of the ultimate purpose of the trouble shooting question, you should be capable of confidently walking through the problem with total composure. In fact, not having a mental breakdown is one victory over the problem. You should be demonstrating how your curiosity and view of the problem as a challenge help motivate you seek the solution.

The "Pressure" (or how much of a geek is this guy/gal?) Question

In particular for companies who use a standard technical test for interviewing recent graduates along with more experienced engineers, the test questions may become progressively more difficult as the test progresses. This allows the interviewer to immediately learn if a candidate is lacking the basic, fundamental knowledge that a new employee is expected to have. If the candidate cannot answer or struggles mightily with the initial questions, there may be little point in continuing to work through the test. This also helps to build a candidates confidence towards the beginning of the testing so that he goes into the more difficult portion of the testing in a state of mind that will help to more accurately demonstrate his knowledge and capabilities.

Even if a candidate easily blows through the initial technical questioning and demonstrates a mastery of the expected knowledge for the position, however, the test may continue on with ever more difficult questioning. This may be used to rank the candidate in comparison to other candidates for positions that see a large number of applicants. More importantly, however, this eventually allows the interviewer to observe the candidate's reaction to difficult problems. A candidate may be able to answer most technical questions with ease, but does he flip out and start slamming his fists against the table in some Incredible Hulk like reenactment as soon as he faces a challenge?

Progressively challenging questioning is not a golden rule, however, so you should not expect that your first technical questions will be easy or panic if the first question is extremely difficult. Some no-nonsense interviewer may want to throw you into the fire pit and see if you survive, especially if his first impression of your technical ability is a positive one.

A standard technical test may have actually been designed for a more advanced position, with smaller portions of it used for more entry level positions, or may include ever more challenging questions to allow observation of your reaction to adversity. Once you are in the hot seat and answering questions, there is little you can do to make up for a lack of knowledge, so do not waste energy and mental resources on things outside of your control (such as wishing you had studied more). Remember your attitude and reaction to facing a challenge are under evaluation just as much as is your technical ability.

How Does Your Brain Work... (Google Wannabe Questions)

How many doctors are there in Boston?

By the way, you have no information or way to get it. Just pretend you are locked in a room with no Internet access or any other source of information and for some bizarre reason you must estimate the number of doctors in Boston. In other words, the "That's the type of information I'd look up" type answers will not suffice in this completely unrealistic scenario...

These types of questions are a personal pet peeve of mine. They generally fall into the "I want to see how your brain works" category. Supposedly, some famous Internet companies ask questions like this, and so they seem to be a fad in the software world, though luckily I haven't seen these types of questions as often in electrical engineering interviews.

The general idea here is to stay cool (don't panic or freak out) and give an answer that demonstrates some sort of logical thought process. First, you may try to estimate the population of Boston based on your existing knowledge of the city. Then, you may try to estimate how many doctors are needed for a given population. A primary care physician can only care for so many patients, for example.

You should also ask specifics, to ensure that 'doctors' in the question is only referring to the medical type. There are a lot of schools in Boston and therefore, many professors with PhDs. After all, if you are stuck in a closed off room trying to make population estimates, you can't make any assumptions.

Luckily, many companies are starting to recognize that such questioning is generally a worthless interview tool. In a New York Times interview with Laszlo Bock, senior vice president of people operations at Google, Block said,

"On the hiring side, we found that brainteasers are a complete waste of time. How many golf balls can you fit into an airplane? How many gas stations in Manhattan? A complete waste of time. They don't predict anything. They serve primarily to make the interviewer feel smart."

Industry Specific Questions

There is likely to be some industry and/or company specific questions that are presented if, for no other reason, they are what your interviewers know and are used to working with on a daily basis. Their expertise will influence the questions they come up with for interviewees. And given its relevance to the company, your understanding of such subjects is obviously more important than your understanding of technologies that may be of little use for their specific position.

It would be hard, if not impossible, to come up with a list of questions for each potential type of electrical engineering job, so no effort has been made to do that here. The main take away from the questions in this book is that you should know the practical fundamentals of electrical engineering. But you should also study up on technology specific to the company you will be interviewing with.

As a new college graduate, a power supply manufacturer is not going to expect you to be a power supply design expert any more than an industrial controls company is going to expect you to be a controls guru, but it is reasonable to expect that their technical portion of the interview is going to be skewed towards those subjects respectively. Doing some prior research into a company's products and technology also demonstrates that you take interest in what they do.